IRINA RODICA RABEJA

WRITINGS II

COMMUNICATING BY ELECTROMAGNETIC WAVES TO 7G /
PREVENTING MALADIES / QUANTUM PHYSICS CONCEPTS

ENGINEERING   SCIENCE

IRINA RODICA RABEJA

WRITINGS II

COMMUNICATING BY ELECTROMAGNETIC WAVES TO 7G /
PREVENTING MALADIES / QUANTUM PHYSICS CONCEPTS

ENGINEERING   SCIENCE

NATIONAL
LIBRARY
OF AUSTRALIA

A catalogue record for this
book is available from the
National Library of Australia

ISBN: 978-0-6486752-6-6

Publisher Irina Rabeja
Sydney Australia
2025

CONTENT

COMMUNICATING BY ELECTROMAGNETIC WAVES

WIRELESS COMMUNICATION

The communication is an intrinsic feature of the human society, communicating between entities or groups of people has been always important in the everyday life of humans. Communication is the activity or process of giving messages or information to others using signals such as speech, body movements or radio signals.

The goal of any communication is to connect and to transmit information.

The word "communicate" comes from the Latin word "communicare" meaning "to share".

The human communication was revolutionized with speech approximately 200,000 years ago, the symbols were developed about 30,000 years ago and the writing appeared about 7,000 years ago.

The communication over a distance, called *telecommunication* - a compound word with the Greek prefix "tele" meaning "far off" - began thousands of years ago with the use of *fire/smoke signals* and *drums/horns* in Africa, America and parts of Asia. Later appeared *mail* in 6th century BC, *pigeon post* in 5th century BC, *hydraulic semaphores* in 4th century BC, *heliographs* in 490 BC, chains of *beacons* in Middle Ages, *maritime flags* in 15th century.

The French inventor **Claude Chappe** (1763-1805) designed the *visual telegraphy system* or *semaphore* between Lille and Paris in 1792 year, which was the first telecommunication system of the industrial age.

The American painter and inventor **Samuel Finley Breese Morse** (1791-1872) developed and patented a *recording electric telegraph* in 1837 year and co-invented with the American machinist and inventor **Alfred Lewis Vail** (1807-1859) the Morse code signalling the alphabet in 1838 year. The first telegrams were sent by Morse on 11 January 1838 across 3 km of wire at Speedwell Ironworks near Morristown New Jersey USA and in 1844 over 71 km of wire from the Capitol in Washington to the old Mt. Clare Depot in Baltimore USA, the latter messaging: WHAT HATH GOD WROUGHT? (Archaic from Bible for "what has God done?")

From then on, commercial telegraphy became successful and popular in America with lines linking in the next decade all the major metropolitan centres on the East Coast.

The *first successful telegraph cable across the Atlantic Ocean* was laid in 1866 year.

Australia was first linked to the rest of the world in October 1872 by a submarine telegraph cable at Darwin.

The American electrical engineer **Elisha Gray** (1835-1901) and the Scottish-American scientist, engineer, inventor **Alexander Graham Bell** (1847-1922) invented the *telephone* same year 1876. *Elisha Gray and Alexander Graham Bell controversy* concerns the question of whether Gray and Bell invented the telephone independently.

The first *commercial telephone services* started between New Haven USA and London UK in years 1878-1879.

The *first telegraph cable across the Pacific Ocean* was completed in 1903 year when finally, the telegraph encircled the world. The cable carried the first message to ever travel around the globe, from USA President Theodore Roosevelt on July 4, 1903. He wished "a happy Independence Day to the USA, its territories and properties". The message took nine minutes to travel worldwide.

"Telegraph" and "telephone" are words with Greek language roots: "graph" meaning "written symbol" and "phone" meaning "sound" beside the prefix "tele" meaning "far off".

The telegraph and the telephone were electrical devices connected by wires or cables.

After them, appeared a variety of experimental techniques for communicating telegraphically "without wires" such as photoelectric and induction telegraphy.

They preceded the "wireless" telegraphy systems, which communicated by "radio waves" and were developed by the Italian inventor **Guglielmo Marconi** beginning in1895 year.

The term "wireless" is derived from the fact that communication may be effected between two points without the aid of wires connecting the points. The term "radio" is derived from the fact that the electromagnetic energy released into space is radiated in all directions.

By 1910 year, the term wireless telegraphy has been largely replaced by the more modern term "radiotelegraphy". In 1912 year, US Navy adopted the term "radio-communication".

The transmission of sound by radio waves or "radiotelephony" began by the 1920s, further making possible radio broadcasting.

"Wireless communication" is the transfer of information between two or more points that are not connected by physical link (an electrical conductor/wire).

The communication by radio waves has been most commonly form of wireless communication. Non-common methods of achieving wireless communications had been considered the light, the magnetic or electric fields or the sound.

The distances encompassed by electromagnetic waves can be short, few meters for television remote control, or very long, millions of kilometres for deep-space radio communications.

The communication by electromagnetic waves encompasses various types of fixed, mobile or portable applications, including two-way radios, cellular telephones, personal digital assistants PDAs - handheld devices that combines computing, telephone/fax, Internet and networking features. Also includes cordless telephones, garage door openers, wireless computer mice, keyboards and headsets, headphones, radio receivers, satellite television, broadcast television and GPS units.

Wireless communications is a broad and dynamic field that has generated great interest and technological awareness over the last few decades.

The use of wireless technology has grown dramatically over the past two decades.

Wireless icon/sign/symbol

The number of wireless devices worldwide has increased exponentially over the past decade, with consumer requirements for faster data rates and longer life devices. The former led to considering for wireless applications electromagnetic waves with frequency above 24 GHz.

More, the Mobile Communications Industry is on the way to connect the World with devices such as smartphones, tablets, wearables and everyday objects, a device revolution that is impacting lifestyles, businesses and institutions. These devices are gaining the ability to communicate wirelessly with each other and with remote locations via sensors, beacons and other data gathering and broadcasting technologies.

That represents the evolutionary role of the device toward becoming a network or relay node. Technologies for communicating, storing, distributing data such as cloud, mesh, device-to-device D2D, peer-to-peer P2P are the key enablers of such a trend.

Inside those devices there are large numbers of sophisticated components such as processors, modems, sensors, chipsets, radios and battery growing in complexity and numbers; driven by cost and performance improvements in digital technologies, their prices lower making easier to incorporate sophisticated multiple functionalities into devices.

Consumer demand for mobile data is exponential and shows no signs of slowing, the consumed mobile data of over 10 Exabytes per month in 2014 year is projected to increase over 70 Exabytes per month by 2020 year.

The Internet of Things IoT is the next big thing in the wireless revolution. It is a natural evolution of the Internet and, as the name suggests, goes beyond the connection of people to the Internet by connecting 'things' such as machines and sensors.

And the use of Low Power Wide Area Networks LPWANs as the communications medium will be the "killer application" that will invade the world creating other millions of applications and implementations over the next decade.

ELECTROMAGNETIC WAVES

The stationary electric charged objects produce an electric field E and the moving electric charged objects produce a magnetic field B.

The combination of the electric field and the magnetic field is the electromagnetic field EMF.

The electric and magnetic fields have around two types of actions:

- non-radiative (near field) action - power is transferred by inductive coupling between coils of wire (most widely used) or by capacitive coupling between metal electrodes.

- radiative (far-field) action - power is transferred by electromagnetic radiation between antennas.

In Physics the electromagnetic radiation EMR refers to the electromagnetic waves EMW of EMF.

The EM waves propagate through space, carrying electromagnetic radiant energy.

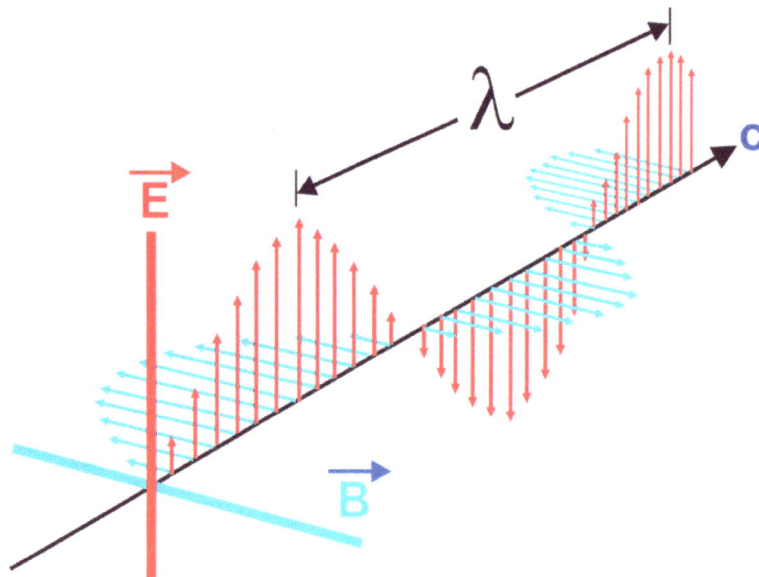

This diagram shows a linearly plane polarized EM wave propagating.
The electric field E is in a vertical plane and the magnetic field B is in a horizontal plane.
The electric and magnetic fields in EM waves are always in phase and at 90 degrees to each other.
EM waves travel in space at the speed of light c = 299,792,458 m/s.
"Plane polarization" of electromagnetic radiation is called the confinement of the electric field vector E or the magnetic field vector B to a given plane along the direction of propagation. Orientation of "linearly" polarized electromagnetic wave is defined by the direction of E. Above EM wave is vertically polarized.

The electromagnetic radiant energy exists everywhere in the universe, it is invisible and works in ways that are still something of a mystery to scientists. In historical terms, the humans harnessed it only some over 100 years ago for communication, which is the transmission of human intelligence and yet the social impact of this new means of communication has been nothing less than phenomenal.

Some say that the historical impact of wireless communications seems as revolutionary for the world as Gutenberg's moveable-type printing press was in the 15th century.

Printing was introduced to Europe by the German blacksmith, goldsmith, engraver, printer, publisher and inventor **Johannes Gensfleisch zur Laden zum Gutenberg** (1400-1468).

His mechanical movable-type printing press started the "printing revolution" and is widely regarded as the most important invention of the second millennium.

Electromagnetic waves are synchronized, self-sustaining double oscillations, one of the electric field E and the other of the magnetic field B, that propagate perpendicular to the direction of propagation and perpendicular to each other.

The electromagnetic waves travel in space at the speed of light c = 299,792,458 metres/second. When travel in an object, the waves are slowed according to that object's magnetic permeability μ and electric permittivity ε.

An electric or magnetic wave is characterized by amplitude, wavelength, frequency and phase. *Amplitude* of wave is basically the height of the wave, marked E for electric wave and B for magnetic wave. Between E and B there is the relation E = cB.

Wavelength of wave, denoted by Greek letter λ, is the distance from one 'peak of wave' to next.

Frequency of wave, denoted by letter f, is the number of 'peaks' per second. Between λ and f there is the relation f = c/λ.

Phase of wave, denoted by Greek letter φ, is the position of a point on the wave.

The manifestation of the electromagnetic field is the *electromagnetic interaction*, one of the four fundamental interactions of Nature, beside gravitation, weak interaction and strong interaction.

In Quantum Physics the electromagnetic radiation EMR consists of photons, the elementary particles responsible for all electromagnetic interactions. Quantum effects give additional sources of EMR, such as the black-body radiation and the transition of electrons in an atom from high to lower energy levels.

The energy of an individual photon is quantized and is greater for photons of higher frequency in conformity to Planck's equation E = hv, where *E* is energy per photon, *v* is frequency of photon and *h* is Planck's constant. As example, a single gamma ray photon could carry over 1 million times the energy of a single photon of visible light.

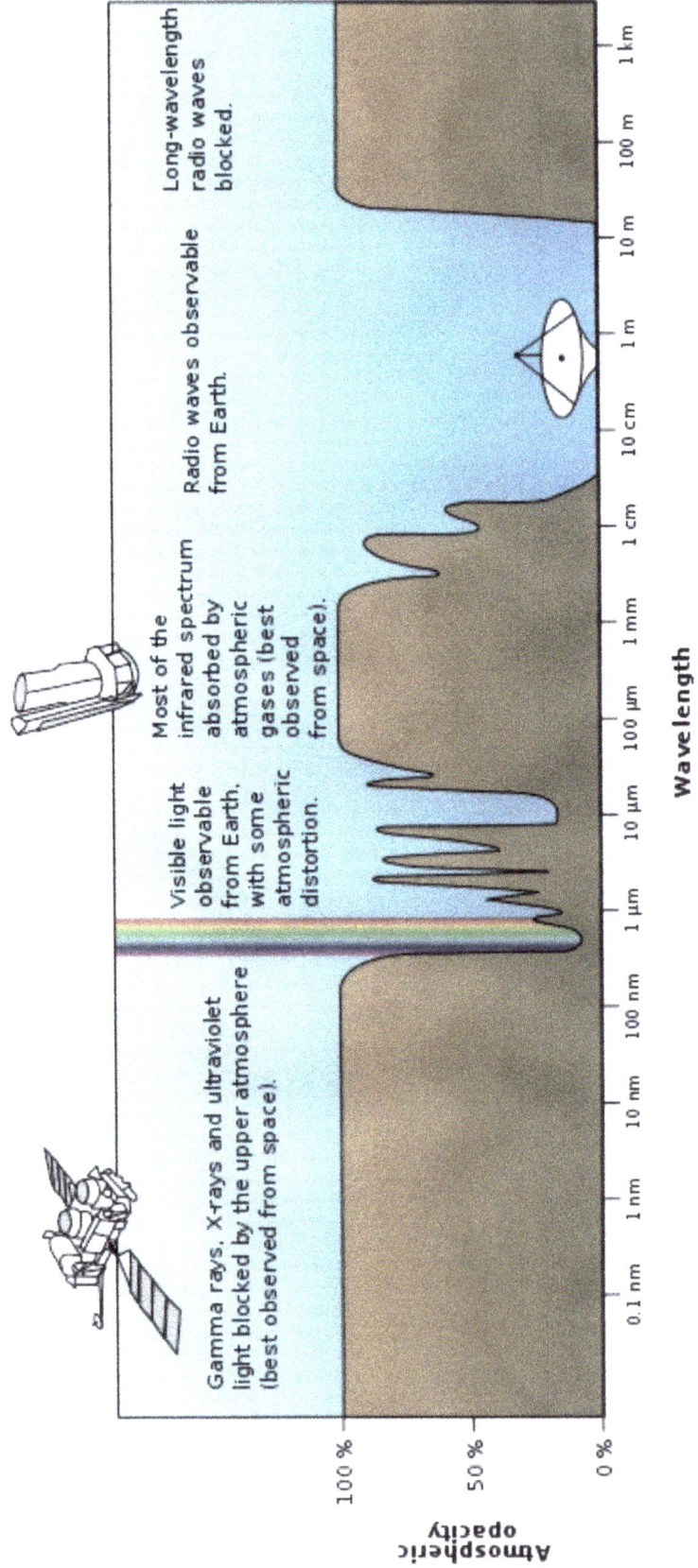

EARTH ATMOSPHERE ELECTROMAGNETIC OPACITY

Gamma rays, X-rays and ultraviolet light blocked by the upper atmosphere (best observed from space).

Visible light observable from Earth, with some atmospheric distortion.

Most of the infrared spectrum absorbed by atmospheric gases (best observed from space).

Radio waves observable from Earth.

Long-wavelength radio waves blocked.

Wavelength

0.1 nm   1 nm   10 nm   100 nm   1 μm   10 μm   100 μm   1 mm   1 cm   10 cm   1 m   10 m   100 m   1 km

Atmospheric opacity

100 %   50 %   0 %

14

In Nature the electromagnetic waves are created by lightning or by celestial bodies.

The regions of the electromagnetic spectrum that pass largely un-attenuated from space through the Earth atmosphere are called "windows". The Earth atmospheric electromagnetic transmittance or opacity is illustrated in the left image.

The range of all possible electromagnetic waves of the electromagnetic radiation, in all their possible lengths or frequencies, is called the *electromagnetic spectrum.*

Across its spectrum, the electromagnetic radiation interacts with the matter in different ways like would be more types of radiation. The spectrum is divided related to these qualitative interaction differences, but the reason invoked is the wavelength, respective the frequency of the electromagnetic waves. So, function of their wavelengths the range of electromagnetic waves is divided in 7 main groups: Radio, Microwave, Infrared, Visible, Ultraviolet, X and Gamma.

The boundaries between some could vary arbitrary in their use for different human activities.

Electromagnetic waves

Name	Wavelength	Frequency
Radio	100,000 km - 0.1 mm	3 Hz - 3 THz
Microwave	1 m - 1 mm	300 MHz - 300 GHz
Infrared	1 mm - 750 nm	300 GHz - 430 THz
Visible	750 nm - 390 nm	430 THz - 770 THz
Ultraviolet	390 nm - 10 nm	770 THz - 30 PHz
X	10 nm - 10 pm	30 PHz - 30 EHz
Gamma	< 10pm	> 30 EHz

Electromagnetic waves of different wavelengths (m) and their scales

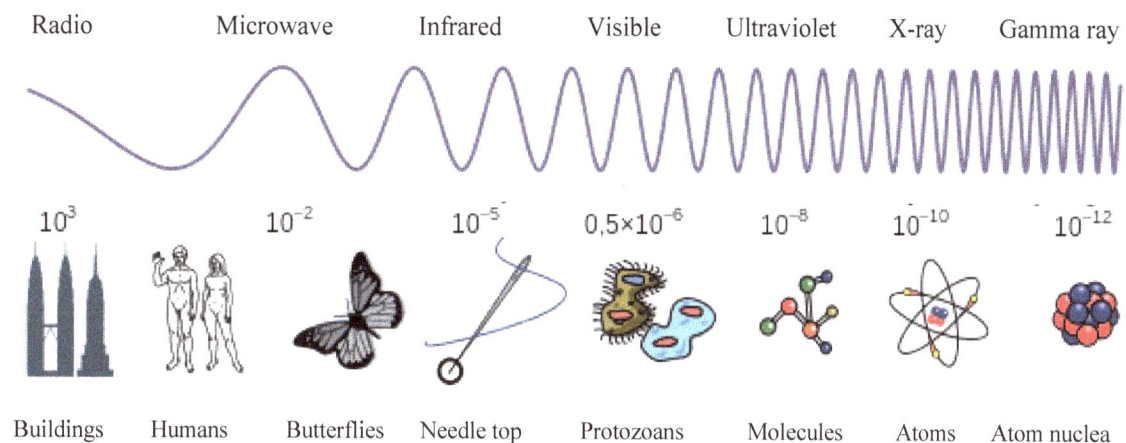

Radio	Microwave	Infrared	Visible	Ultraviolet	X-ray	Gamma ray

| 10^3 | 10^{-2} | 10^{-5} | $0,5 \times 10^{-6}$ | 10^{-8} | 10^{-10} | 10^{-12} |

| Buildings | Humans | Butterflies | Needle top | Protozoans | Molecules | Atoms | Atom nuclea |

Electromagnetic waves of different frequencies experience differences in reflection, refraction, polarization, diffraction, absorption and how move in free space and over the surface of Earth. The electromagnetic waves are used by humans for different purposes, function of their different frequencies. They are used in communication, radar, radio astronomy, navigation, heating and power application, spectroscopy (interaction between matter and electromagnetic radiation), atomic absorption spectroscopy AAS (determination of around 70 elements), high-resolution imaging, medicine.

The communication by electromagnetic waves is also called "wireless communication" using mainly the radio waves and the microwaves.

Radio waves RW are the electromagnetic waves in the range of 0.1 mm-100,000 km for wavelength and in the range of 3 Hz-3 THz for frequency, including waves used for communication or radar signals.

Their main interaction with the matter is the collective oscillation of charge carriers in bulk material (plasma oscillation). An example is the oscillatory travels of the electrons in an antenna. International Telecommunication Union ITU is a specialized agency of UN responsible for issues concerning information/communication technologies. Radio Regulations RR of ITU regulates on law of nations scale radio-communication services and utilisation of radio frequencies.

ITU divides the radio waves spectrum in 12 bands, denoted by symbols, each beginning at a wavelength that is a power of 10 and covering a decade of wavelengths (or frequencies).

That originated with a recommendation of the fourth meeting of the Consultative Committee on International Radio CCIR held in Bucharest, Romania in 1937 year and was approved by the International Radio Conference held in Atlantic City, New Jersey, USA in 1947 year.

B.C. Fleming-Williams in 1942 year suggested to give each band also a number, which is the logarithm of the approximate geometric mean of the upper and lower band limits in Hz, for example the approximate geometric mean of Band 6 is 10^6Hz.

ITU Radio Bands Symbols	ELF	SLF	ULF	VLF	LF	MF	HF	VHF	UHF	SHF	EHF	THF
ITU Radio Bands Numbers	1	2	3	4	5	6	7	8	9	10	11	12

Function of their frequency, the radio waves have different propagation characteristics in the Earth's atmosphere: the very high frequency ones travel on the line of sight, bend and reflect very little, the high frequency ones called sky waves can reflect on ionosphere and return to earth beyond horizon and the low frequency ones called ground waves have low attenuation, suitable for long distance communication because diffract around obstacles.

Radio waves created by men are generated by different devices and are used for fixed and mobile radio communication, broadcasting, radar, navigation systems, communications satellites, computer networks and others.

Like all other electromagnetic waves, the radio waves travel through space at the speed of light.

Radio waves

Band Number	Band Symbol	Frequency Range	Wavelength Range
1	ELF	3-30 Hz	10000-100000 km
2	SLF	30-300 Hz	1000-10000 km
3	ULF	300-3000 Hz	100-1000 km
4	VLF	3-30 kHz	10-100km
5	LF	30-300 kHz	1-10km
6	MF	300-3 MHz	100-1000m
7	HF	3-30 MHz	10-100m
8	VHF	30-300 MHz	1-10m
9	UHF	300-3000 MHz	10-100cm
10	SHF	3-30 GHz	1-10cm
11	EHF	30-300 GHz	1-10mm
12	THF	300 -3000 GHz	0.1-1mm

THF = Tera frequencies (far-infrared waves)

EHF = Extremely high frequency (microwaves)

SHF = Super-high frequency (microwaves)

UHF = Ultra-high frequency (radio waves)

VHF = Very high frequency (radio waves)

HF = High frequency (radio waves)

MF = Medium frequency (radio waves)

LF = Low frequency (radio waves)

VLF = Very low frequency (radio waves)

ULF = Ultra-low frequency (radio waves)

SLF = Super-low frequency (radio waves)

ELF = Extremely low frequency (radio waves)

Millimeter-wave mmW and *submillimeter-wavs* sub-mmW are radio spectrum bands in the range 0.3-3 mm / 100-1000 GHz and *terahertz wave* THz is radio spectrum band in the range 0.1-1mm / 300-3000 GHz. They are receiving increasing attention bringing distinct benefits, such as wider bandwidth, higher spatial and temporal resolution, more compact antennas and reusability of frequencies.

The sparsely used electromagnetic spectrum between 100 GHz and 1000 GHz commonly known as millimetre-wave and sub-mmW regions, can be conquered by the current rapid development of electronic circuits and subsystems beyond 100 GHz enabled by improvements in high-frequency semiconductor technology (using gallium nitride GaN and indium phosphide InP and new packaging techniques).

In recent years mmW and THz applications for spectroscopy and imaging, astronomy and environmental/atmospheric study and monitoring have grown more and more rapid as push forward techniques. The mmW/THz imaging is viewed as a safe, low-cost alternative to usual techniques for biological, security and health-sciences applications.

Close, applications in areas of high-speed wireless communication using carriers at 38 GHz, 68 GHz, 81-86 GHz and 92-95 GHz for defence, security and space science have been increasing rapidly.

Microwaves MW are defined as the electromagnetic waves with wavelength in the range 1 mm-1 m and frequency in the range 300 MHz-300 GHz.

Their main interactions with the matter are plasma oscillation and molecular rotation.

They belong to upper frequencies range of radio waves including UHF, SHF, EHF but are defined separately because of their special applications. Most common applications of microwaves are in the 1- 40 GHz range.

The prefix "micro-" is not for micro- range, it indicates that they have shorter wavelengths than the waves used in radio broadcasting, which is the unidirectional wireless transmission of radio waves for large audience.

Different sources define different frequency ranges for microwaves. In general, the boundaries between ultra-high frequency radio waves, microwaves, terahertz radiation and far infrared light vary arbitrary in their use for different activities.

All warm objects emit microwave radiation function of their temperature, so microwave radiometers can be used to measure the temperature.

The galaxies, their stars emit microwave radiation, which is studied by radio astronomers using cosmic microwaves receivers called radio telescopes.

The cosmic **m**icrowave **b**ackground **r**adiation CMBR is considered "relic radiation" from the inception of the universe. Due to expansion followed by cooling of the universe, the originally high-frequency electromagnetic radiation has been shifted to microwave-frequency radiation. Sensitive radio telescopes can detect the faint, omnidirectional, background glow CMBR, which is not associated with any star, galaxy or other celestial body.

Man-made sources of microwaves use specialized vacuum tubes and field effect transistors and tunnel, Gunn, IMPATT diodes.

Maser (acronym for **m**icrowave **a**mplification by **s**timulated **e**mission of **r**adiation) is the device that produces and amplifies coherent microwaves.

Microwaves travel by line-of-sight paths. Therefore, on the surface of the Earth microwave communication links are limited by the visual horizon to about 48-64 km.

Microwaves are easily focused in narrow beams, allow broad bandwidth, need small antenna size for transmitters, receivers or transceivers.

Microwaves are the principal means that transmit data, TV and telephone speech between ground stations and satellites. They are used extensively for point-to-point telecommunications, in radar technology and microwave ovens.

The microwaves spectrum is divided in 13 bands: L, S, C, X, Ku ,K, Ka, Q, V, E, W, F, D

Microwaves

Band Symbol	Frequency Range	Wavelength Range
L	1-2 GHz	15-30 cm
S	2-4 GHz	7.5-15 cm
C	4-8 GHz	3.75-7.5 cm
X	8-12 GHz	25-37.5 mm
Ku	12-18 GHz	16.7-25 mm
K	18-26.5 GHz	11.3-16.7 mm
Ka	26.5-40 GHz	5.0-11.3 mm
Q	33-50 GHz	6.0-9.0 mm
V	50-75 GHz	4.0-6.0 mm
E	60-90 GHz	3.3-5 mm
W	75-110 GHz	2.7-4.0 mm
F	90-140 GHz	2.1-3.3 mm
D	110-170 GHz	1.8-2.7 mm

L is used for military telemetry, GPS, mobile phones GSM, amateur radio

S is used for weather radar, surface ship radar, some communication satellites

C is used for long distance radio telecommunications

X is used for satellite communications, radar, terrestrial broadband, space communications, amateur radio

Ku is used for satellite communications

K is used for radar, satellite communications, astronomical observations

Ka is used for satellite communications

Q is used for satellite communications, terrestrial microwave communications, radio astronomy, automotive radar

V is used for millimetre wave radar research and other scientific research

E is used for UHF transmissions,

W is used for satellite communications, millimetre wave radar research, military radar targeting and tracking applications, non-military applications, automotive radar

F is used for SHF transmissions: radio astronomy, microwave devices/communications, wireless LAN, most modern radars, communication satellites, satellite direct broadcasting television DBSTV, amateur radio

D is used for EHF transmissions: radio astronomy, high frequency microwave radio relay, microwave remote sensing, amateur radio, directed-energy weapon, millimetre wave scanner

Infrared waves IRW are electromagnetic waves with wavelength in the range 750 nm-1 mm and frequency in the range 300 GHz-430 THz.

Their main interactions with matter are molecular vibration and plasma oscillation (only metals). Everything with a temperature above 5 degrees Kelvin [or -450 °F or -268 °C] emits IR radiation. It is, beside convection and conduction, the way heat is transferred from one place to another. The infrared electromagnetic radiation has industrial, scientific, medical and short-ranged wireless communication applications, is important in remote sensing, spectroscopy and weather forecasting.

Visible waves VW or *Light* are the part of the electromagnetic spectrum that is visible to the human eye. A typical human eye will respond to electromagnetic waves with wavelength in the range of 390-750 nm and corresponding frequency in the range of 430-770 THz.

Their main interactions with matter are molecular electron excitation (including pigment molecules found in the human retina) and plasma oscillations (in metals only).

The visible spectrum does not contain all the colours that the human eye can distinguish; un-saturated colours such as pink or magenta are not present since they are a mix of electromagnetic waves of different frequencies. The colours containing electromagnetic waves of only one frequency or wavelength are called "pure" colours or "spectral" colours.

The future promises a green technology where office lighting will have a double-duty, will be a source of light and a wireless transmission source. The idea is that a Light Emitted Diode LED can vary its intensity so quickly that a human eye cannot see it, but a photo detector can detect it. That is the Visible Light Communication VLC or Light Fidelity LiFi, which is one of the solutions to the problem about the lack of radio spectrum.

LiFi refers to the high-speed, bidirectional and networked wireless communications using light to provide a seamless wireless user experience much like traditional mobile communications. It advances the visible light communication VLC, which was first introduced by the Japanese professor **Nakagawa** of Keio University. At Keio Techno-Mall in December 2009, there were demos of visible light communication and the latest research results were presented.

LiFi offers secure and safe wireless communications in a globally unlicensed spectrum that repurposes the energy used for lighting to provide wireless data. LiFi is a complete mobile communication solution to augment the fifth generation of cellular phone called 5G and beyond. LiFi technology is expected to be a key part of the future 5G Systems.

How to integrate LiFi into existing WiFi connections is a compelling problem but its solution will promise more data to customers. How to modulate, multiplex and handover LiFi data is under debate in engineering circles, possibly an **o**rthogonal **f**requency **d**ivision **m**ultiplexing OFDM. Radio is the very utilised part of the electromagnetic spectrum but visible/light and infrared have been underutilised parts of spectrum and both are unlicensed.

The visible/light spectrum alone stretches from approximately 430 THz to 770 THz that is over hundred times the bandwidth of the radio electromagnetic spectrum.

Ultraviolet waves UVW have shorter wavelength and higher frequency than the visible waves/light. Their wavelength is in the range of 10 nm-390 nm and their frequency is in the range of 770 THz-30 PHz.

Their main interaction with matter is the excitation of molecular and atomic valence electrons, including ejection of the electrons (photoelectric effect).

They are produced by Sun or artificial sources.

They are invisible to humans but near UV radiation is visible to some insects and birds.

UV lamps sterilise the air in operating theatres, surgical equipment, drugs, food and in suitable doses cause the human body to produce vitamin D, treat vitamin D deficiency or skin disorders.

X waves XW or *X rays* are called the waves of the electromagnetic radiation with the wavelength ranging in 10pm-10 nm and frequency in the range 30 PHz-30 EHz.

Their main interaction with matter is the excitation and ejection of core atomic electrons called Compton scattering for elements of low atomic numbers.

X rays are produced when the electrons strike a metal target.

They are the kind of rays that could travel through solid wood or flesh and, because they impressed photographic plates as light, they can yield photographs of people's bones and are used in diagnostic radiology.

Gamma waves GW or *Gamma rays* or *γ rays* denoted by Greek letter gamma γ, are called the waves of the electromagnetic radiation with wavelength less than 10 pm (10×10^{-12} m) and frequency above 30 EHz (30×10^{18} Hz) and therefore are composed of high energy photons.

Their main interactions with matter are the energetic ejection of core electrons in heavy elements called Compton scattering for all atomic numbers, the excitation of atomic nuclei including dissociation of nuclei and for high-energy rays the creation of particle-antiparticle pairs or a shower of high-energy particles-antiparticles.

Gamma rays are ionizing radiation, therefore dangerous to the people's health.

Gamma radiation is generated by gamma decay of radionuclides/radioisotopes (unstable atoms), atmospheric interaction with cosmic ray particles, lightning strikes, terrestrial gamma-ray flashes TGF caused by intense electric fields above or inside thunderstorms (last 0.2-3.5 ms with energy up to 20 MeV), astronomical processes producing high-energy electrons causing deceleration radiation, nuclear fusion in stars, collapse of hyper-novae stars (give the most powerful bursts of gamma rays).

Gamma rays are used in medicine in radioactive tracers, to sterilise medical equipment, to treat internal organs, to kill cancer cells.

Electromagnetic spectrum is the range of all possible electromagnetic waves of the electromagnetic radiation, in all their possible wavelengths or frequencies. It is a naturally occurring resource much like air or water.

The word "spectrum" was introduced in the year 1666 by the genial English mathematician and Nature scientist **Isaac Newton** (1643-1727) when he directed a ray from Sun through a prism and saw that on the wall of his room appeared the colours of the rainbow. The prism decomposed the light in a row of colours entering lightly one in another from red, through orange, yellow, green, blue, indigo to violet. To name the multicolour band that appeared like by magic on the wall of room, Newton took from the Latin language the word "spectrum" meaning "ghost appearance" or "phantom" (plural "spectra" or "spectrums").

In general "spectrum" defines a condition that can vary continuously, is not limited to a set of values.

The range of all possible electromagnetic waves of the electromagnetic radiation are presented further by their wavelength λ on a decimal logarithmic scale and the corresponding frequency f, resulted in conformity with the formula $f = c/\lambda$, also on a decimal logarithmic scale.

Communicating by electromagnetic waves implies the use of the electromagnetic spectrum and it is here where the possibilities for radio communications begin and end.

With the exploding popularity of all things wireless, the radio spectrum has become a scarce commodity in many countries.

Radio spectrum, which is finite, must accommodate mobile phone calls and data traffic that are increasing at an unprecedented rate. Globally, the traffic on mobile broadband systems has grown so fast that current levels already exceed predictions made in the 2010 year for the 2020 year according to Huawei, a global information and communications technology provider in China.

The efficient use of spectrum required the coordinated development of standards, which in turn contributed to the development of technologies that relied on spectrum use.

Spectrum management is the process of regulating the use of radio frequencies to promote efficient use because of the increasing number of spectrum uses in air-broadcasting, commercial services to the public, government, industrial, scientific, medical services.

Cognitive radio CR is one technology under development that could allow spectrum to be used more efficiently. A CR transceiver scans for unused spectrum bands and changes its transmission and reception parameters to different frequencies during heavy data loads without interruption. It also can listen for interference on busy channels and calculate a way to reduce it.

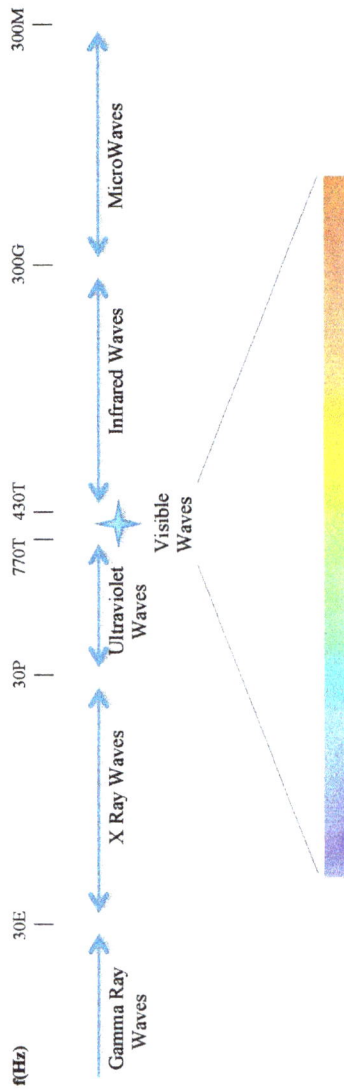

ELECTROMAGNETIC SPECTRUM

λ(m)   1p   10p   100p   1n   10n   100n   1μ   10μ   100μ   1m   3m 10m 30m 0.1 0.3 1 3 10 30 100 300 1k 3k 10k 30k 100k 1000k 10^7

f(Hz)   300E 30E 3E 0.3E/300P 30P 3P 0.77P 0.3P/300T 30T 3T 0.3T/300G 30G 3G 0.3G/300M 30M 3M 0.3M/300k 30k 3k 0.3k/300

f(Hz)   430T 770T

f(Hz)   30E   30P   300G   300M

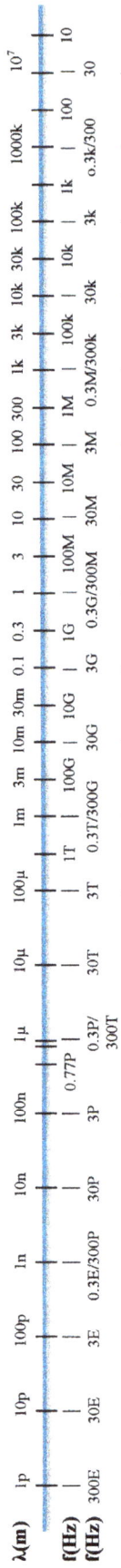

ITU Radio Band symbols: THF EHF SHF UHF VHF HF MF LF VLF ULF SLF ELF

ITU Radio Band numbers: 12 11 10 9 8 7 6 5 4 3 2 1

Radio Waves

| mm W |

Gamma Ray Waves   X Ray Waves   Ultraviolet Waves   Visible Waves   Infrared Waves   MicroWaves

For wireless communication to be effective was necessary driving government intervention and international coordination.

International Telecommunication Union ITU coordinates internationally the use of radio electromagnetic spectrum, managing interference and setting global standards.

ITU was formed in 1865 year in Paris, at the International Telegraph Convention. Its first regulations were put in place at the 1906 Berlin International Radiotelegraph Conference.

In 1947 year, ITU became a United Nations specialized agency for information and communication technologies ICTs. It is based on public-private partnership and currently has a membership of 193 countries and almost 800 private-sector entities and academic institutions.

ITU has headquarters in Geneva, Switzerland and 12 regional and area offices around the world.

ITU coordinates the shared global use of the radio spectrum (used in modern technology, particularly in telecommunication), promotes international cooperation in assigning satellite orbits, works to improve telecommunication infrastructure in the developing world, assists in the development and coordination of worldwide technical standards.

ITU deals with interference by requiring member countries to respect notification and registration procedures whenever they plan to assign an electromagnetic wave of a particular frequency to a certain use like a radio station or a new satellite. Spectrum users are divided into primary and secondary, with primary users protected from interference from secondary users but not vice versa.

DISCOVERY OF ELECTROMAGNETIC WAVES

Apart from the discoveries and inventions associated with electricity generation, the discovery of electromagnetic waves and the inventions that followed have been responsible for major changes in the world. Without the communications by electromagnetic waves many of those changes would not have happened.

The light or the visible electromagnetic radiation is the only part of the electromagnetic spectrum known by humans for most of the history.

The study of light began in ancient Greece, regarding light reflection and light refraction.

In the 16th century appeared conflicting light theories considering light either wave or particle.

The British-German astronomer and composer **Sir William Hershel** (1738-1822) in 1800 year discovered the "infrared radiation" when studying the temperature of different colours and observing that the highest temperature was beyond red.

The German chemist, physicist and philosopher **Johann Wilhelm Ritter** (1776-1810) in 1801 year discovered beyond the violet light rays the "chemical rays", invisible rays inducing chemical reactions. They were renamed later "ultraviolet c radiation".

The English scientist **Michael Faraday** (1791–1867) introduced the concept of *field* in physics to describe his discovery of electromagnetic induction in 1821 year.

Michael Faraday had little formal education, but was one of the most influential scientists, considered by the historians of science the best experimentalist in the history of science.

It was by his research on the magnetic field around a conductor carrying a direct electric current that Faraday *established the basis for the concept of the electromagnetic field* in physics.

He showed that the electromagnetic field generated by charged bodies and magnets extended in the empty space around by lines of flux, used too as way to visualize electric and magnetic fields.

Faraday also established that there was an underlying relationship between magnetism and light by discovering in 1845 year the magneto-optical effect: the plane of vibration of a linearly polarized light beam incident on a glass piece rotates when a magnetic field was applied in the direction of beam propagation.

Faraday invented the first electric motor, the first electrical transformer, the first electric generator and the first dynamo, so Faraday can be called without any doubt, **the *father of electrical engineering.*** However, the scientists of his time widely rejected his ideas.

In his honour, at the 1881 International Congress of Electricians in Paris was introduced officially the name "farad" symbol F for the SI unit of electrical capacitance - the body ability to store electrical charge.

The Scottish mathematical physicist **James Clerk Maxwell** (1831-1879) understood Faraday's ideas and taking in consideration the Faraday, Ampere, Gauss laws did the notable achievement in 1865 year to formulate the *classical theory of electromagnetic radiation* in his publication "A Dynamical Theory of the Electromagnetic Field" bringing together for the first time electricity, magnetism and light as manifestations of the same phenomenon, electromagnetic waves traveling in space at light speed.

Maxwell's equations for electromagnetism producing an unified model of electromagnetism represent one of the greatest advances in physics. The unified model of electromagnetism has been called "the second great unification in physics" after the first one realised by Isaac Newton.

The first major unification in physics was Isaac Newton's realization that the same force that caused an apple to fall at the Earth's surface, the gravity, was also responsible for holding the Moon in orbit about the Earth and this universal force would also act between the planets and the Sun, providing a common explanation for both terrestrial and astronomical phenomena.

Maxwell's equations are a set of partial differential equations that together with the Lorentz force law form the foundation of classical electromagnetism, classical optics and electric circuits.

MICHAEL FARADAY

Lorentz force is the action of electric and magnetic fields on a point electric charge: a particle of electric charge q moving with the velocity **v** in the presence of an electric field **E** and a magnetic field **B** experiences a force $\mathbf{F} = q(\mathbf{E} + \mathbf{v} \times \mathbf{B})$.

The English self-taught electrical engineer, mathematician and physicist **FSR Oliver Heaviside** (1850-1925) reformulated the Maxwell's original twenty field equations in more convenient four equations. Their formulation and meaning are:

$\nabla\mathbf{E} = q/\varepsilon_0$	The electric flux leaving a volume is proportional to the electric charge q inside
$\nabla\mathbf{B} = 0$	There are no magnetic monopoles; the total magnetic flux through a closed surface is zero
$\nabla \times \mathbf{E} = -\,\partial\mathbf{B}/\partial t$	The electric field **E** induced in a closed circuit is proportional to the rate of change of the magnetic flux it encloses
$\nabla \times \mathbf{B} = \mu_0 (J + \varepsilon_0\,\partial\mathbf{E}/\partial t)$	The magnetic field **B** induced around a closed loop is proportional to the electric current plus the displacement current (rate of change of electric field) it encloses

Nabla or *del* symbol ∇ represents in mathematics the vector differential operator.

ε_0 is the vacuum permittivity or the permittivity of free space, measure of the amount of resistance encountered when forming an electric field E in classical vacuum and its value is:

$\varepsilon_0 = 8.8541878176\times10^{-12}$ F/m

μ_0 is the vacuum permeability or the permeability of free space, measure of the amount of resistance encountered when forming a magnetic field B in classical vacuum and its value is:

$\mu_0 = 4\pi \times 10^{-7}$ H·m^{-1} ≈ $1.2566370614...\times10^{-6}$ H·m^{-1} or N·A^{-2}

James Clerk Maxwell is considered the leading theoretical physicist of the 19th century. International Electrotechnical Commission in 1930 year honoured James Maxwell by naming the CGS unit of magnetic flux the "maxwell" symbol Mx.

The dramatic confirmation of Maxwell's theory that light itself is an electromagnetic wave challenged experimentalists to generate and detect electromagnetic radiation.

The first electromagnetic transmission-reception was made by the British-American inventor, experimenter and music professor **David Edward Hughes** (1831-1900) in 1879 year, but was not conclusively proven to be transmission through the air of electromagnetic waves or merely electromagnetic induction.

JAMES MAXWELL

Experiments done by the German physicist **Heinrich Rudolf Hertz** (1857 –1894), who engineered instruments to transmit and receive electromagnetic waves pulses, used in 1886 year experimental procedures that excluded all other known wireless phenomena.

To emit radiation, to build a "transmitter", Hertz used a high voltage induction coil, an original form of capacitor called condenser or Leyden jar and a "spark-gap" device connected in series. He caused a spark discharge between the spark-gap poles (2 cm radius) oscillating at a frequency determined by the values of the capacitor and the induction coil.

To detect the emitted radiation, to build a "receiver", Hertz used 1 mm thick copper wire, bent into a 7.5 cm diameter circle, at one end with a small brass sphere and at the other end with a pointer directed to sphere at a distance controlled by a screw mechanism (hundredths of mm). The receiver was designed so that oscillating back-forth electric current in the "receiver" wire would have the frequency of the "transmitter" current. The presence of oscillating charge in the receiver would be signalled by sparks across tiny gap between pointer and sphere.

In other experiments, Hertz established beyond any doubt that light is a form of electromagnetic radiation by measuring the speed of electromagnetic radiation and founding it to be the same as the light's speed and showing that the radio waves' reflection and refraction were same as light's. Hertz's proof of the existence of airborne electromagnetic waves led to an explosion of experiments with this new form of electromagnetic radiation called *Hertzian waves* until approximately 1910 year when the term *radio waves* became current.

Heinrich Rudolf Hertz's experiments caused large interest in radio waves research.

Inventions that followed used the radio waves to transfer information through space.

Eventually, successful wireless telegraph, audio radio and later television were developed commercially.

Hertz died young in 1894 year before his discoveries were implemented into a practical form, unaware of the practical importance of his radio wave experiment about which he said:

"It's of no use whatsoever... just an experiment that proves Maestro Maxwell was right - we just have these mysterious electromagnetic waves that we cannot see with the naked eye. But they are there." About the ramifications of his discoveries Hertz said: "Nothing, I guess."

International Electrotechnical Commission in 1930 year honoured Heinrich Hertz by naming the SI unit of frequency, the one cycle per second unit, the "hertz" symbol Hz.

HEINRICH HERTZ

The German physicist **Wilhelm Conrad Roentgen** (1945-1923) on 8 November 1895 discovered a kind of rays that could travel through solid wood or flesh and impress photographic plates as light. After some considerable investigation, he understood that he discovered a new kind of rays and he named them "X" (unknown).

The French physicist **Antoine Henri Becquerel** (1852-1908) in 1896 year discovered that uranium salts produce rays which he called "uranium rays", proved to be a blend of three kinds of radiation of which one is "gamma" γ.

FIRST EXPERIMENTS

The Scottish philologist, astronomer, inventor and author **James Bowman Lindsay** (1799-1862*) introduced the concept of wireless telegraphy.*
In 1832 year, Lindsay gave a classroom demonstration to his students, of wireless telegraphy via conductive water. By 1854 he was able to demonstrate wireless transmission, using water as the transmission medium from Dundee city to Woodhaven village in Scotland situated at 3km distance. He obtained a patent for his system of wireless telegraphy through water.

The American businessman and inventor **Thomas Alva Edison** (1847-1931), described as "America greatest inventor" in 1880s *patented an electromagnetic induction system he called grasshopper telegraphy*, which allowed telegraphic signals to jump the short distance between a running train and telegraph wires running parallel to the tracks. The system was successful technically but the train travellers were little interested in an on-board telegraph service.
To develop wireless telegraph systems were used both electrostatic and electromagnetic induction. But they had very limited commercial application.
Edison has no formal education but was a prolific inventor with 1093 patents in US and other many in UK, France and Germany. The phonograph, the motion picture camera, the stock ticker, the mechanical vote recorder, the electrical light bulb, the battery of an electric car, electric power generation and distribution, first industrial research laboratory are part of his inventions.

By experiments between the 1884 -1886 years, the Italian physicist and inventor **Temistocle Calzecchi-Onesti** (1853-1922) at Fermo in Italy demonstrated that *iron powder in insulated tube conducts electric current when under electromagnetic radiation* generally known as *radio waves*.
This primitive device was developed later in 1890 to become the first practical "radio detector" by the French physicist, professor and inventor **Edouard Eugene Desire Branly** (1844-1940).

JAMES LINDSAY

The British physicist and writer **Sir Oliver Joseph Lodge** (1851-1940) used it as "coherer" in 1893 year. The *coherer* is considered a primitive form of radio signal detector used in the first radio receivers during the wireless telegraphy era at the beginning of the 20th century. It was used also in the first commercially successful radio transmission system between 1894 and 1896 years by the Italian electrical engineer and inventor Guglielmo Marconi.

The Welsh electrical engineer and inventor **Sir William Henry Preece** (1834-1913) was the most successful *creator of an electromagnetic induction telegraph system* beginning with tests across the Bristol Channel in 1892, able to telegraph across gaps of about five kilometres. However, his induction system required extensive lengths (km) of antenna wires, at both the sending and receiving ends.

The Brazilian Catholic priest inventor **Father Roberto Landell de Moura** (1861-1928) between 1890 and 1894 years *conducted wireless transmissions in telegraphy and telephony* over distances of up to 8 kilometres, demonstrating a radio broadcast of the human voice on 3 June 1900. In 1904 year, he obtained three patents for: Wave Transmitter, Wireless Telephone, Wireless Telegraph

The Serbian-American inventor, electrical engineer, mechanical engineer, physicist and futurist **Nikola Tesla** (1856-1943) after 1890 year *experimented to transmit electric power through inductive and capacitive coupling,* which is *wirelessly,* using spark-excited radio frequency resonant transformers now called *Tesla coils.*

Tesla did *public demonstrations of wireless transmission,* lighting Geissler tubes and incandescent bulbs from across a stage in his attempt to develop a wireless lighting system based on near field inductive and capacitive coupling.

Tesla found that by using a receiver LC circuit in resonance with a transmitter LC circuit he can increase the distance at which he can light a lamp. *This resonant inductive coupling is used nowadays in short-range wireless power systems.*

Tesla called attention to the fact that by taking his electric oscillator, grounding one side of it and connecting the other to an insulated body of large surface, it should be possible to transmit electric oscillations to a great distance and to communicate intelligence in this way to other oscillators in sympathetic resonance therewith. This was going far toward the invention of radio-telegraphy.

In 1891 year, he *developed various electrical generators,* devices that convert mechanical energy to electrical energy in the form of alternating current AC with the frequency of 15000 cycles per second. Alternating current AC is an electric current, which periodically reverses direction whereas a direct current DC flows only in one direction.

THOMAS EDISON

On 3 February 1892 Nicola Tesla presented at the Institution of Electrical Engineers of London the lecture "Experiments with Alternate Currents of High Potential and High Frequency", when he *suggested that messages could be also transmitted without wires* and *the telephony could be rendered practicable across the Atlantic Ocean*. The lecture was repeated on 4 February at the Royal Institution London and on 19 February at the Societe Francaise de Physique in Paris.

In February 1893 Tesla presented the lecture " On Light and Other High Frequency Phenomena" at the Franklin Institute, Philadelphia and in March 1893 he repeated the presentation before the National Electric Light Association, St. Louis. On 25 August 1893 before the International Electrical Congress Tesla delivered the lecture "Mechanical and Electrical Oscillators".

In his experiments Tesla *exhibited transmission and radiation of radio frequency energy or electromagnetic power transfer & proposed to be used for telecommunication of information too*.

The English-American electrical engineer and editor **Thomas Commerford Martin** (1856-1924) published in 1893 year the book " The Inventions, Researches and Writings of Nikola Tesla" detailing the work of Nikola Tesla which contained coupled oscillation circuits each having capacitors & inductors in series.

In 1897 year, the *Tesla method* was described in New York, USA as what is *known today as the Wireless Power Transfer WPT*. Wireless power techniques mainly fall into two categories:

* non-radiative (near field) - power is transferred by inductive or capacitive coupling.

Tesla demonstrated that for the first time when he lit Geissler tubes and even incandescent light bulbs from across a stage.

* radiative (far-field or power beaming) - power is transferred at long distances by beams of electromagnetic radiation as microwaves or laser beams aimed to receiver.

In 1898, Tesla developed a radio-controlled robotic boat driving the boat remotely around the waters of Manhattan from a set of controls at Madison Square Garden.

Tesla proposed a "World Wireless System" that was to broadcast both information and power worldwide, including the building of more than thirty transmission-reception stations, large high-voltage wireless power stations near major population centres of the world.

By the end of 1900 year the financier J. P. Morgan agreed to fund a pilot project, the Wardenclyffe project, able to transmit messages, telephony and even facsimile images across the Atlantic to England and to ships at sea.

Morgan was interested financially in shares in the company and half of all the patent income.

But Tesla decision in July 1901 to include wireless power transmission met Morgan's refusal to fund the changes and by 1904 year the investment stopped.

NICOLA TESLA

Nikola Tesla demonstrating wireless transmission of power and high frequency energy   at Columbia College, New York, in 1891.

The two metal sheets were connected to his Tesla coil oscillator, which applied a high voltage oscillating at radio frequency.

The electric field ionized the gas in the long partially-evacuated Geissler tubes he is holding (similar to modern neon lights), causing them to emit light without wires.

N. Tesla's Wardenclyffe Wireless Station/Wardenclyffe Tower/Tesla Tower

Located in Shoreham, Long Island, New York - seen in 1904 year

Since today global demand for energy, two Russian physicists **Leonid** and **Sergey Plehanov** intend to realize their project for wireless energy transmission once proposed by the 20th century scientist Nikola Tesla. They say that solar panels and an upgraded Tesla tower could solve the present global energy problem.

Tesla displayed his scientific imagination and created widespread interest in his brilliant demonstrations. By doing that he stimulated the scientific imagination of many others.

Tesla obtained hundreds of patents worldwide (more than 1,200 patents) for his inventions, but there are also inventions, not accounted for, hidden in archives or without patent protection.

There has been stated that Tesla actually discovered alternating electric current AC, produced the first AC induction motor, invented radio (preceded Marconi with few years), discovered the arc light and broadcasted the first television signals. Tesla's electromagnetic receivers were more responsive than coherers used later by others. He too worked on wireless remote-control devices.

Some say more, that Nikola Tesla was the genius who created the modern world: electric cars, radio, the bladeless turbine, wireless communication, spark plugs, fluorescent lighting, the induction motor, the telephone repeater, the rotating magnetic field principle, the poly-phase alternating current system, alternating current power transmission, the Tesla Coil transformer. And in addition, the principles he discovered and the mechanisms he invented led to TV, MRI, X-ray machines, radio telescopes, radar. He envisioned the smart phone and wireless Internet.

He was one of the few who combined the brain of a calculator with the imagination of a scientist. Tesla used mathematics to arrive at many of the concepts that he turned into inventions.

He did his engineering drawings in detail without ever measuring a line and the machined parts from his drawings fit perfectly. But very few of his subtle mathematical calculations of his designs exist in journals and in conformity with the saying "publish or perish" he is considered today an engineer and an inventor but not a scientist. But his genius is recognized.

Some sources say that the names of Thomas Alva Edison and Nikola Tesla were announced by the Swedish Academy to share the 1912 Nobel Prize in Physics but ultimately the prize was awarded to Swedish inventor engineer Nils Gustaf Dalen for the invention "automatic regulators for use in conjunction with gas accumulators for illuminating lighthouses and buoys".

In 1960 year, the General Conference on Weights and Measures honoured Nicola Tesla by naming the SI unit of magnetic flux density, the "tesla" symbol T.

The Italian inventor and electrical engineer **First Marquis of Marconi Guglielmo Giovanni Maria Marconi** (1874-1937) *opened the way for modern wireless communications in the year 1895 by transmitting the 3 dots Morse code for the letter "S" over a distance of 3 km using electromagnetic waves.*

Marconi read about the experiments of Heinrich Hertz and also read about Nicola Tesla's work. He understood that radio waves could be used for wireless communications.

Marconi's early apparatus was a development of Hertz's laboratory apparatus into a system designed for communications purposes.

In July 1896, Guglielmo Marconi presented his invention as a new method of telegraphy to the attention of William Henry Preece, then engineer-in-chief to the British Government Telegraph Service, interested himself in the development of wireless telegraphy by the inductive-conductive method. Preece stated at the Royal Institution in London that Marconi had invented a new relay, which had high sensitivity and delicacy.

By 1897 year, Guglielmo Marconi *did more demonstrations with a radio system for signalling over long distances*. In 1899 year were transmitted messages across the English Channel.

In 1901 year he received in St. John's Newfoundland, Canada telegraphic signals across the Atlantic Ocean sent from his wireless station in Poldhu, Cornwall, England UK at ~3200 km.

At the turn of 20th century, Guglielmo Marconi developed the first apparatus for long distance radio communication.

By 1910 year different various wireless systems were referred as "radio".

Marconi was an entrepreneur, businessman and founder of The Wireless Telegraph & Signal Company in the United Kingdom in 1897 year, renamed Wireless Telegraph Trading Signal Company in 1900 year. Marconi's real contributions were engineering and commercial.

For contributions in development of wireless telegraphy Guglielmo Marconi was awarded the **1909 Nobel Prize for Physics.**

The Russian physicist **Aleksander Stepanovich Popov** (1859-1905), acclaimed in his homeland and some eastern European countries as the inventor of radio, in 1897 year succeeded to transmit radio waves over a distance of 5 km. He did not pursue their use for communications but used them to study thunderstorms.

The German physicist and inventor **Karl Ferdinand Braun** (1850-1918) *discovered* in 1874 year that can rectify alternating current by a point-contact semiconductor, called *cat's whisker detector* or *crystal detector - the first type of semiconductor diode*, an electronic component consisting of a thin wire that lightly touches a crystal of semiconducting mineral (usually galena).

In 1897 Braun built the first cathode-ray tube CRT and the CRT oscilloscope.

He was captivated by wireless telegraphy in 1898 year, when he joined the wireless pioneers.

He *introduced a closed tuned circuit in the generating part of a wireless transmitter and separated it from antenna by an inductive coupling. At receiver he used crystal detectors.*

GUGLIELMO MARCONI

Braun's British patent on tuning was used by Marconi in many of his tuning patents.

Braun experimented first wireless at the University of Strasbourg.

In 1899 year, he experimented on the shore of the North Sea and in 1900 year, radio telegraphy signals were exchanged regularly with the island of Helgoland over a distance of 62 km.

Light vessels in the river Elbe and a coast station at Cuxhaven independent town commenced a regular radio telegraph service.

Braun contributed significantly to development technology for radio and television.

Ferdinand Braun and Guglielmo Marconi shared the **1909 Nobel Prize for Physics** for contributions in development of wireless telegraphy.

In the early 20th century, the Slovak inventor, architect, botanist, painter Catholic priest **Josef Murgas** (1864-1929) *did revolutionary work in wireless telegraphy.* In 1905 year, his company "Universal Ether Telegraph Co." organized a public test of its transmitting/receiving facilities. Josef Murgas obtained more patents related to his inventions in the wireless telegraphy.

TECHNOLOGY EVENTS

Fleming valve or *Fleming oscillation valve*, a "vacuum tube" or "thermionic diode" was invented in the year 1904 by the British electrical engineer and physicist **John Ambrose Fleming** (1849-1945). It was a tube with 2 electrodes, the first "diode" whose purpose was to conduct electric current only in one direction, used as *detector* for early radio receivers in electromagnetic wireless telegraphy. Later was used as *rectifier*, device that converts alternating current AC into direct current DC in the power supplies.

The American inventor, self-described "father of Radio" **Lee De Forest** (1873-1961) invented in the year 1906 the *Audion,* a vacuum tube with 3 electrodes actually the Fleming valve with a third electrode, the first "triode" that permits control the strength of tube electric current without consuming appreciable energy. It was used to build the first amplifying radio receivers and electronic oscillators.

In 1920 year, the commercial radio was established in U.S. with radio station WWJ in Detroit and radio station KDKA in Pittsburgh.

In January 1926 year the Scottish engineer and innovator **John Logie Baird** (1888-1946) demonstrated the first working mechanical television system. He invented the first colour television system and the first purely electronic colour television picture tube.

In 1929 year was established the British Broadcasting Corporation BBC.

In 1940 year the American, Bell Labs researcher **George Robert Stibitz** (1904-1995) recognized as one of the fathers of first digital computer, was able to transmit problems using "tele-printers", devices for communicating text over telegraph lines, to his "Complex Number Calculator" in New York and received the computed results back at Dartmouth College in New Hampshire. This was a *configuration of a centralized computer with remote terminals*, which remained popular till 1950 year.

In 1946 year was heralded in press the "Giant Brain" a new computer with a computing speed 1000 times faster than electro-mechanical machines called the *Electronic Numerical Integrator And Computer* ENIAC. It was the first electronic digital general-purpose computer, able to simulate any "Turing machine".

Turing machine is an abstract machine that manipulates symbols according to certain rules. Given an algorithm, a Turing machine can be constructed to simulate the logic of that algorithm. The English computer scientist, mathematician, logician, cryptanalyst and theoretical biologist **OBE FRS Alan Mathison Turing** (1912-1954) invented the Turing machine. He is considered the father of theoretical computer science and artificial intelligence.

In 1946 Motorola company in conjunction with the Bell System operated the *first commercial portable/mobile telephone service* Mobile Telephone System MTS in the USA.

In 1960s were launched for the first time the communication satellites. These first satellites could only handle 240 voice circuits. They have become essential in places where would be impossible to communicate by any other method.

In 1960 year appeared a new computer communication method and technology, called *Packet switching*, enabling data to be divided in packets and transmitted through different paths or nodes to one final point.

In December 1969 year emerged a 4 Node network linking University of California Los Angeles, Stanford Research Institute, University of Utah, University of California Santa Barbara. This network extended under the name *Advanced Research Projects Agency Network* ARPANET was based on concepts and designs of American engineer and computer scientist **Leonard Kleinrock,** Polish-American engineer **Paul Baran** (1936-2011), Welsh computer scientist **OBE FRS Donald Watts Davies** (1924-2000), American scientist **Lawrence Roberts**. As the project evolved, were developed protocols allowing multiple separate networks to join into a network of networks.

The American **Steve Crocker** in 1969 year invented the "Request For Comments" RFC to help record unofficial comments on the development of ARPANET.

In 1973 year, ARPANET added a non-US node, the NORSAR Project of Norway, followed by a node in London. In 1981 year, ARPANET had 213 nodes in USA.

In 1981 year RFC introduced "Internet Protocol v4" IPv4 & "Transmission Control Protocol" TCP creating for ARPANET the protocol suite *Transmission Control Protocol/Internet Protocol* TCP/IP that much of Internet relies today. The Americans electrical engineer **Robert Elliot Kahn** and mathematician **Vinton Gray Cerf** with concepts of the French engineer **Louis Pouzin** invented TCP/IP communications protocols.

In 1982 year, RFC introduced the *Simple Mail Transfer Protocol* SMTP, a standard for electronic mail (e-mail) transmission. It was updated in 2008 year and is the protocol in widespread use today. ARPANET would eventually merge with other networks to form the modern *INTERNET* and many of the protocols the Internet relies today were specified through this process.

The Jewish-American engineer, computer scientist and professor at University of Hawaii **Norman Manuel Abramson** in 1960 year *developed the first computer network communicating by electromagnetic waves / the first wireless computer communication network* called ALOHAnet or ALOHA system or simply ALOHA using low-cost ham-like radios.

In the Hawaiian language the word "aloha" means affection/peace/compassion/mercy. ALOHA had 7 computers deployed over 4 islands and a central computer on the Oahu Island. ALOHA communicating without using phone lines, became operational in June 1971 year when demonstrated public the first wireless packet data network.

The first commercial automated cellular network (first generation cellular phones G1) was launched by the Nippon Telegraph and Telephone NTT corporation in Japan in 1979 year, followed by the launch of Nordic Mobile Telephone NMT system in Denmark, Finland, Norway and Sweden in 1981 year. They used analogue technology and offered only voice services.

Early laptop/portable computers appeared in 1980s with the first generation of wireless data modems developed by amateur communication groups.

In 1985 year, the Federal Communication Commission FCC, America's telecom regulator, transferred parts at 900MHz, 2.4GHz and 5,8GHz from ISM bands to communications entrepreneurs. ISM bands are portions of the radio spectrum reserved internationally for the use of radio frequency RF energy for industrial, scientific and medical purpose.

Much later in 1997, inspired by the wire-line networking standard Ethernet was realised a common wireless standard. The equipment based on the new wireless standard operated under the name "Wi-Fi" in the 2.4 GHz and 5.8GHz bands, allowing data transfer at 2Mb/s and using spread spectrum technologies.

In 1989 year ,the Voyager 2 spacecraft sent pictures of planet Neptune down to Earth laying the basis for satellite communication.

NORMAN ABRAMSON

Also, in 1989 year "World Wide Web" WWW was invented by the English scientist **Sir Timothy John Berners-Lee** when he proposed an information management system.

In 1996 year, RFC introduced a protocol that made possible the hyperlinked Internet, called *Hypertext Transfer Protocol* HTTP, an application protocol for distributed, collaborative, hypermedia information systems including graphics, audio, video, plain text and hyperlinks. HTTP is the foundation of data communication for the WWW, an information space where documents and other web resources are identified by Uniform Resource Locators URLs, are interlinked by Hypertext links and can be accessed by Internet.

The Internet access became wide spread using the old telephone and television networks.

Apple company's laptop computers offered *first wireless connectivity in 1999 year* introducing Wi-Fi as an option under the name AirPort.

Wi-Fi is a short-range networking technology, an introduction in what will be possible with future wireless technology.

From now wireless communication has developed into a key element of modern society. Wireless communication has revolutionised the way the societies function, from satellite transmission, radio and television broadcasting, to the now ubiquitous mobile telephone.

In 1999 year, the Japanese firm NTT DoCoMo released the first "smartphones". *Smartphone* is a cell mobile phone with an advanced mobile operating system that combines features of personal computer operating system with features useful for mobile use. Smartphones became widespread in the late 2000s.

In 2007 Apple's iPhone was the first touchscreen smartphone to gain mass-market adoption. Most of smartphones produced from 2012 onward have high-speed, motion sensors and mobile payment features. 1 billion smartphones were in use worldwide in 2012. Global smartphone sales surpassed the sales figures for regular cell phones in early 2013.

Bluetooth technology was unveiled in year 1999 but only in 2000 year the manufacturers began to adopt it in mobile phones and computers.

In 2001 year Apple company launched the *iPod,* which together with Apple's iTunes software, was the technology that really transformed the way people listened to music. The device's sleek design made it desirable to own and with its large internal storage capacity was no longer necessary to carry around CDs or cassette tapes.

In 2003 year, the instant messaging application/computer program that provides online text message and video chat services called *Skype* has transformed the way people communicate across borders, speak and even video chat over WiFi. Initially only available as a desktop client, over time Skype was launched on mobile.

Nissan company manufactured a compact five-door hatchback electric car, a *leading environmentally-friendly affordable family car* LEAF, introduced in Japan and USA in Dec 2010 and in European countries and Canada in 2011 year.

IBM Watson is an artificially intelligent computer system capable of answering questions asked in an ordinary language. In year 2011, it competed on the American quiz show *Jeopardy* and beat the two greatest human champions. It represented an important milestone in the development of artificial intelligence AI - a field that has been progressing rapidly with innovations like the motion sensor Microsoft Kinect and the ordinary language voice command Apple Siri. In another breakthrough for AI, in 2016 year the artificial neural network *DeepMind* beat the world human champion at *Go*, an abstract strategy board game for two players in which the aim is to surround more territory than the opponent.

Google began testing self-driving vehicles in California in 2012 year with the intention to make them available by 2017. The cars have a top speed of 25 miles/hour and are designed to be perpetually in motion. If the concept becomes successful, it is thought that driverless cars could transform the way we move around cities in the future.

TERMS, CONCEPTS, TECHNICS & TECHNOLOGIES

Signal is a series of radio waves, light waves, electrical impulses.

Antenna or *aerial* is a rod, wire or other structure by which signals are transmitted or received as part of a radio (electromagnetic) transmission or receiving system, an essential component of all equipment using radio.

Wireless is used usually to describe communications in which electromagnetic waves carry an information signal over part/entire communication path from transmission to reception.

Carrier of message/information is the electromagnetic wave.

Message or *information* is the speech/voice, images/photos, business data.
To be transmitted, the message/information is converted into an electrical form, an electric analogue or digital/discrete signal by suitable converters and added to the carrier. An analogue signal can be represented as a sum of simple sine waves of different frequencies in a frequency range w, the Fourier series. The modern technologies prefer the signal in digital/discrete form so the analogue signals are transformed in digital ones by "sampling".

Sampling is the method by which an analogue signal with the highest frequency w is transformed in discrete samples or digital signal providing that the sampling rate exceeds 2w samples in conformity with the Shannon-Nyquist Theorem.

Modulation is the addition of message/information (in electrical form) to a carrier.

A device called *modulator* performs the modulation.

The information/message in electrical form is called the *modulating signal*.

The carrier wave has usually a much higher frequency than the modulating signal.

The modulation is done by varying/modulating one or more features of the carrier, for example: amplitude modulation, frequency modulation, phase modulation

Demodulation is the extraction of message from *modulated carrier*.

A device called *demodulator* performs the demodulation.

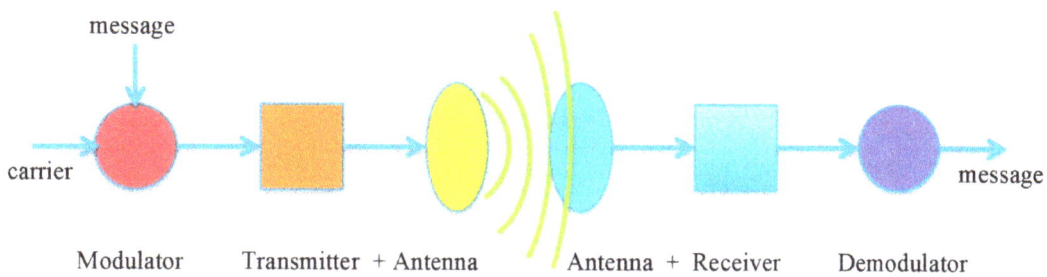

Communicating by electromagnetic waves

Bandwidth is a central concept in many fields, including electronics, information theory, digital communications, radio communications, signal processing and spectroscopy expressing the difference between the upper frequency and lower frequency in a continuous set of signal frequencies and is measured by *hertz* symbol Hz.

Communication channel is a concept in telecommunication and computer networking, referring to physical transmission medium/space/cable/wire of a signal or a multiplexed signal. The range of frequency of the electromagnetic waves that can be transmitted through a channel is called the *channel bandwidth* B and is measured by *hertz* symbol Hz.

Capacity is a concept used in connection with the communication channel. In electrical engineering, computer science and information theory, the *channel capacity* C is the upper bound of the rate at which information can be reliably transmitted over a communications channel and is measured by *bits per second* b/s.

The capacity C of a transmission channel with an **a**dditive **w**hite Gaussian **n**oise AWGN and signal to **n**oise ratio S/N, providing the channel bandwidth B, is stated by the Shannon-Hartley Theorem as $C = B \log_2(1+S/N)$ where:

B is the bandwidth given in Hz.

S/N is the signal-to-noise ratio expressed as a power ratio and is dimensionless.

C is the capacity measured in bits per second b/s if the logarithm is taken in base 2, or nats per second nat/s if the natural logarithm in base e=2.71828 is used.

Wider the bandwidth and lower the noise, greater the capacity of the communication channel.

Wider/*broader* the bandwidth of channel, greater is the information-carrying capacity of channel.

Data transfer rate DTR is the term for the amount of digital data that is moved from one place to another in a given time, that is the speed of travel of a given amount of data from one place to another, measured in *bits per second* b/s.

Greater the bandwidth of a given path, higher is the data/information transfer rate/speed.

Examples of data transfer rates are:

To the Internet a typical low-speed connection may be 33.6 kb/s.

On Ethernet local area networks, data transfer can be as fast as 10 Mb/s.

Network switches are planned to transfer data in the Tb range (1Tb=10^{12} bits).

Data transfer time between the microprocessor or RAM and devices such as the hard disk and CD-ROM player is usually measured in milliseconds.

In computers, data transfer is often measured in *bytes per second* B/s (1 byte=8 bits).

The highest data transfer rate to date is 14 Tb/s over a single optical fibre, reported by Japan's Nippon Telegraph and Telephone company NTT DoComo in 2006 year.

Broadband is a relative term, understood according to its context. Originally the word "broadband" used as "uncountable" was a concept in telecommunications meaning a wide band of electromagnetic waves frequencies, a technical term. Now "broadband" is a wide bandwidth data transmission with the ability to simultaneously transport multiple signals and traffic types. The medium can be coaxial cable, optical fiber, wire, wireless.

In consequence *broadband* is a high-capacity transmission technique using a wide range of frequencies, which enables a large number of messages to be communicated simultaneously.

In the context of Internet access, *broadband* is used to mean any high-speed Internet access that is always on and faster than the original Internet access technology, the traditional dial-up access, which was limited to 56 kb/s. The term became popularized through the 1990s as a marketing term for Internet access that was faster than the dial-up access.

The USA Federal Communications Commission FCC re-defined the *broadband,* to mean download speeds of at least 25 Mb/s and upload speeds of at least 3 Mb/s.

According to some standards, *broadband* means: having instantaneous bandwidths greater than 1 MHz and supporting data rates greater than about 1.5 Mb/s

Information capacity is a concept in Web hosting service, expressing the amount of data transferred to/from the website/server within a prescribed period of time and is measured in *gigabytes/time-period*.

To express the maximum amount of data-transfer each month it is used the term *monthly data transfer* which is measured in *gigabytes/month*.

Spectral efficiency, spectrum efficiency or *bandwidth efficiency* refers to the information rate that can be transmitted over a given bandwidth in a specific communication system.

It is measured in b/s/Hz.

Coverage is the concept indicating a geographic area where transceiver station can communicate. Broadcasters and telecommunications companies, in order to indicate to users the transceiver station's intended service area, produce *coverage maps* frequently.

The coverage depends on different factors: orography (mountains), buildings, technology, radio frequency, sensitivity & transmit efficiency of consumer equipment.

Electromagnetic waves of certain frequencies provide better regional coverage, while waves of other frequencies penetrate better through obstacles, such as buildings in cities.

The ability of mobile phones to connect to base station depends on the signal strength, which can be boosted by higher power transmissions, better and taller antennas or alternative solutions like in-building pico-cells. Normal Macro-Cell signals need to be boosted to pass through buildings, a problem in designing networks for large metropolitan areas with modern skyscrapers, hence the current drive for small-cells and micro-cells and pico-cells.

Signals also do not travel deep underground, so other transmission solutions are necessary for mobile phone coverage into underground parking garages and subways.

Seamless mobility is one of the paramount demands of the portable/mobile telephone users which gives the users access to mobile content with automatic switching between protocols, networks, communication channels, offers seamless access and connectivity across personal, local and wide area networks and allows the users to roam among home, office, car, hotspots, airports, campus and beyond without interruption.

Shared medium or *shared channel* is a medium/channel that serves more than one user at the same time, so the same communication medium/channel can be used by more transmitters.

Multiplexing or *muxing* is the concept by which multiple analogue or digital signals are sent at the same time over the same communication channel in a form of single complex signal: for example several telephone calls can use one wire.

The receiver recovers the separate signals by the method called *demultiplexing* or *demuxing*.

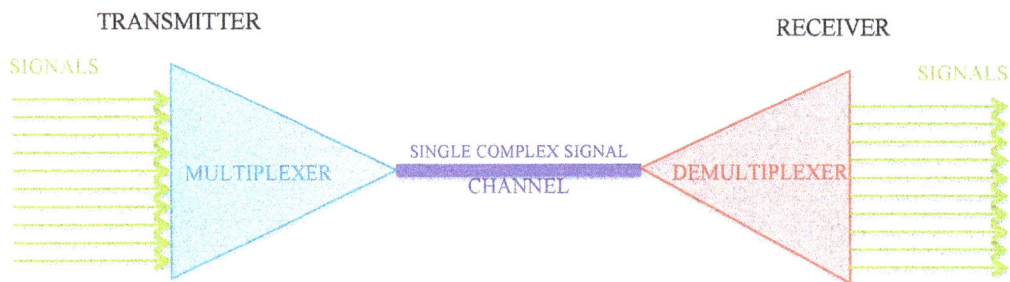

Multiple data signals are multiplexed as a single complex signal before transmitted over the communication channel and de-multiplexed at channel end

The American author and inventor **Major General George Owen Squier** (1865-1934) is credited with the development of the telephone carrier multiplexing in 1910 year.

Multiplexing is based on a multiple access protocol and control mechanism known as *media access control* MAC.

Media access control deals with issues such as addressing, assigning multiplex channels to different users and avoiding collisions. MAC is a component of the TCP/IP model.

The multiplexing is divided in more categories:

Space Division Multiplexing SDM - controls the radiated energy for each user in space.

SDM is achieved with multiple antenna elements forming a phased array antenna.

Examples are multiple-input and multiple-output MIMO, single-input and multiple-output SIMO and multiple-input and single-output MISO multiplexing.

Polarisation Division Multiplexing PDM - is achieved using the polarisation of electromagnetic radiation for separation.

Frequency Division Multiplexing FDM - combines several signals in one by allocating to each signal a unique frequency band.

Time Division Multiplexing TDM - uses time to separate the signals, sending each signal through channel at different moments in time.

Code Division Multiplexing CDM - each user is assigned a special code sequence or "signature" to modulate its message signal, so all users are allowed to transmit over the same channel simultaneously.

FDM, TDM and CDM multiplexing are realised by the multiple access techniques: frequency division multiple access FDMA, time division multiple access TDMA, code division multiple access CDMA.

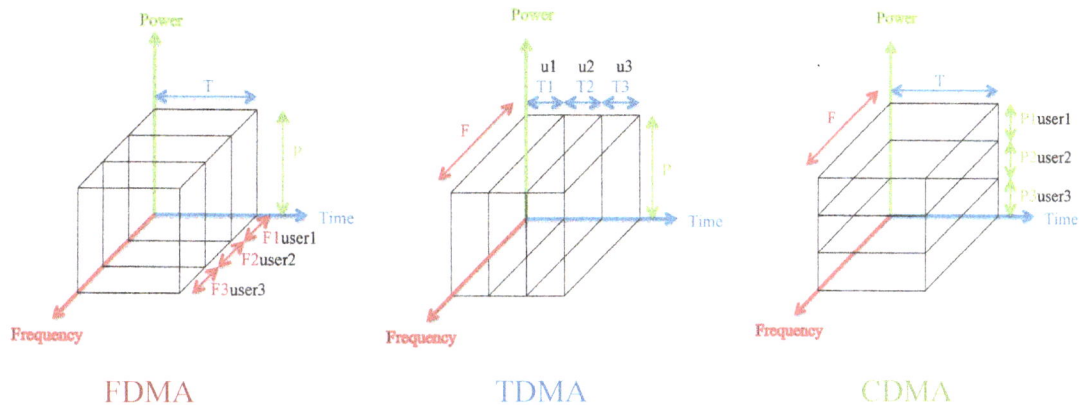

FDMA   TDMA   CDMA

In FDMA the radio spectrum is divided into a set of frequency slots and each user is assigned a frequency slot to transmit. In TDMA several users transmit at the same frequency but in different time slots. In CDMA the users transmit at same frequency and time but the signals are modulated with high bandwidth spreading waveforms, called signature waveforms or codes, which have very low cross-correlation (similarity). In theory, CDMA, TDMA and FDMA have the same spectral efficiency but practically, each has its own challenges - power control in case of CDMA, timing in case of TDMA and frequency generation/filtering in case of FDMA.

Wireless is used in connection with information transmitters, information receivers or information transceivers (combined transmitters & receivers), computer networks, network terminals, remote controls, microwaves, infrared light, laser light, visible light, acoustic energy which transfer information without the use of wires.

Wireless connections may involve point to point communication, point to multipoint communication, broadcasting, cellular networks or other networks.

Wireless technology - technology that uses electromagnetic waves to transmit and receive data - it is used for long-range communications that are impossible or impractical to implement with the use of wires, to span a distance beyond the capabilities of typical cabling or financially impractical, to provide backup communications link in case of network failure, to link portable or temporary workstations, to remotely connect mobile users or networks.

Wireless networks include cell phone networks, computer networks as local area networks LANs or wide area networks WANs, wireless sensor networks, satellite communication networks and terrestrial microwave networks.

Wireless networking could be the method by which communications networks (collection

53

of linked communication points or nodes), enterprise installations and homes avoid the costly process of introducing cables into a building or between various equipment locations.

Wireless communication networks are generally implemented and administered using the electromagnetic waves.

A wireless communication network has to offer *coverage, capacity, data rate/speed, continuity, security, quality of service, cost efficiency*. For example, wireless wide-area networks are limited by coverage in low traffic areas and by capacity in high traffic areas.

Wireless communication can develop a wide range of *services* like broadcasting services, portable/mobile communications services for voice and data including maritime, aeronautical, airplanes, land communications; it can also develop satellite, amateur radio, military, radio astronomy, meteorology and science services.

Bluetooth is a wireless technology standard for exchanging data over short distances using the radio electromagnetic waves UHF in the ISM band from 2.4 to 2.485 GHz for fixed and mobile devices and personal area networks PANs.

ZigBee is a specification for a suite of high-level communication protocols used to create personal area networks with small, low-power digital radios for home automation, medical device data collection and other low-power low-bandwidth needs, designed for small scale projects which need wireless connection. The technology defined by the ZigBee specification is intended to be simpler and less expensive than Bluetooth or Wi-Fi.

Z-Wave is a wireless communications protocol used primarily for residential control and automation market like wirelessly control lighting, security systems, home cinema, garage, spa, swimming pool, automated window, HVAC (**h**eating, **v**entilation, **a**ir **c**onditioning) and home access. Like other protocols and systems aimed at home and office automation market, a Z-Wave automation system can be controlled via the Internet.

Wi-Fi is a wireless technology that allows a computer, laptop, mobile phone or tablet device to connect at high speed to the Internet. The rise of the Internet and the consumer desire to be connected to it at all times drove deployment on high-speed data communications on Wi-Fi. The wireless industry organization Wi-Fi Alliance considers more than 15 billion Wi-Fi enabled devices in 2016 year. Wi-Fi technology is cheap and embedded in a growing number of devices, making it the technology of choice for the foreseeable future.

WiMAX is described as a standards-based technology enabling the delivery of wireless broadband access as an alternative to cable at 2-11 GHz. In practical terms, WiMAX would operate similar to Wi-Fi but at higher speeds, greater distances and for larger number of users.

Initially designed to provide 30 to 40 Mb/s data rates, with the 2011 update WiMAX provides up to 1 Gb/s for fixed stations. It could potentially erase the suburban and rural blackout areas that have no broadband Internet access because phone and cable companies did not yet run necessary wires to those remote locations.

Global area network GAN is proposed as the final step in the area network range. GAN is a type of interconnection of terminals that does not have a geographical limitation and can connect computers from various countries. It is a network regrouping several computers and LANs together in a bigger net.

MOBILE PHONE EVOLUTION

A *mobile phone* is a *portable telephone* that can make/receive calls over an electromagnetic wave link from any location in a certain geographical area.

A mobile phone is an electronic telecommunications device, a wireless handheld device that allows users to make calls and send text messages, among other features.

The mobile phone is one of the greatest successes of human communication.

The earliest generation of mobile phones could only make and receive calls, now they offer a long list of capabilities that bring an even longer list of challenges for service providers.

Today's mobile phones are packed with many additional features, such as web browsers, games, cameras, video players and even navigational systems.

The history of portable or mobile telephones can be broken into 8 periods: pre-cellular, first generation of cellular phone, second generation of cellular phone, third generation of cellular phone, fourth generation of cellular phone, fifth generation of cellular phone, sixth generation of cellular phone and seventh generation of cellular phone.

1. Pre-cellular period was the period of Mobile Radio Telephone systems, preceding the first generation of cellular telephones. These early mobile telephone systems are distinguished from earlier closed radiotelephone systems because are a commercial service, part of the public switched telephone network PSTN with own telephone numbers and not part of closed networks of police or taxi radio systems.

Motorola Company in conjunction with the Bell System (the system of companies led by the Bell Telephone Company) operated the first commercial mobile telephone service called "Mobile Telephone System" MTS in the USA in 1946 year, followed by "Improved Mobile Telephone Service" IMTS with the first automatic dialling in 1964 year.

West Germany launched "A-Netz" in 1952 year, UK launched "System1" in 1959 year, USSR launched the first automatic mobile system in Europe "Altai" in 1965 year,

Norway opened "Televerket" in 1966 year, Finland launched "Autoradiopuhelin" ARP in 1971 year, West Germany launched in 1972 year the second commercial mobile phone network "B-Netz" which was the first not requiring human operators to connect calls. These mobile telephones were placed in vehicles as cars or tracks but briefcase models there were also. Only few people were able to use this device because only 25 channels were available and they could connect to local telephone network only if it was in the range of 20 km.

In USA in parallel to IMTS was a competing mobile telephone technology called "Radio Common Carrier" RCC, which was operated by private companies and individuals.

The introduction of *cellular technology* for portable/mobile phones, greatly expanded the efficiency of the electromagnetic spectrum use. The mobile phone became the cell phone. Cellular technology has multiple low-power transmitters of 100W or less.

Because the range of such transmitters is small, a large geographic area is divided into small geographic areas or cells each served by its own antenna. Different users in different, but non-adjacent, cells are able to use the same electromagnetic wave for a call without interference. The hexagonal shape of cells provides for equidistant antennas at distance d function of radius R of hexagon d=R$\sqrt{3}$. As cells become smaller, antennas move from the tops of tall buildings or hills, to the tops of small buildings or sides of large buildings, finally to lampposts where they form microcells.

A simplified cellular system is illustrated further.

The mobile phone MP is wirelessly connected to the Base Station BS supported by the Base Station Controller BSC. Traditional voice circuit VC is supported through a Mobile Switch Centre MSC both directly and in connection to a public switched Telephone Network PSTN. The BSC can also be connected to an IP Gateway/Router to support packet data services.

In next figure:

MP = Mobile Phone

BS = Base Station

BSC = Base Station Controller

MSC = Mobile Switching Centre

VC = Voice Circuit

PSTN = Public Switched telephone Network

IP GW = IP Gateway

IPDN = Internet Packet Data Network

OS = Operator Service

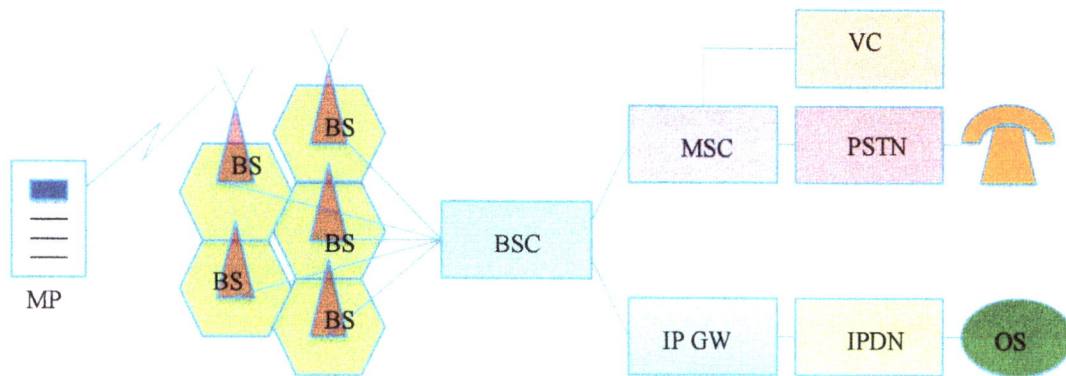

Basic Wireless Communication Cellular System

The history of cellular telephones covers the next 7 periods of the history of mobile telephones under the name of "generations". Each generation G has new features, which differentiate it from the previous one and generally refers to a change of system, technology, speed and frequency.

2. *First Generation* of cellular phone /1G/ was launched in Japan in 1979 year and developed around the world using different, incompatible analogue technologies. While Scandinavia developed the "Nordic Mobile Telephone" NMT system, in USA there was the "Advanced Mobile Phone System" AMPS, UK had the "Total Access Communications System" TACS, Germany developed "C-Netz". The result was a wide range of largely incompatible systems particularly in Europe, although AMPS system was used throughout USA.
A 7-cell reuse pattern was adopted for AMPS, which used analogue technology for frequency modulation FM in frequency band 800-900 MHz with FDM multiplexing, data rates 1-2.4 kb/s and offered only voice/speech services.

3. *Second Generation* of cellular phone /2G/ developed in 1980-1990 years reversed in digital the earlier analogue mobile experience 1G by using the new digital technology.
Second generation digital systems can be classified by their multiple access techniques as FDMA, TDMA or CDMA. In practice TDMA and CDMA are combined with FDMA.
In Europe, a common standard was adopted, partly due to government intervention, called "Group Special Mobile" GSM, which was the first 2G system.
Under the auspices of the European Technical Standards Institute ETSI, GSM was standardized in 1990 year with the new name "Global System for Mobile Communication", which described the protocols for 2G networks and first deployed in Finland in December 1991.

The standardized GSM could allow a mobile phone to be used outside the range of its home network, connecting to other available cell network by full international roaming, automatic location services, common encryption and relatively high quality audio.

GSM is now the most widely used 2G system worldwide, in more than 200 countries.

In contrast, a variety of incompatible 2G standards developed in the United States of America. The failure of USA to adopt a common 2G standard, with the associated benefits in terms of roaming and switching of handsets, led to first generation AMPS system remaining the most popular mobile technology in USA throughout the 1990s.

2G provides primary benefits as: digital encryption, spectrum significantly more efficient used and data services such as text, picture and multimedia messages. It uses the frequency bands 850-1900 MHz (GSM) and 825-849 MHz (CDMA) and data speeds/rates 14 kb/s - 64 kb/s.

Generation /2.5G/ developed in 2000-2003 years introduced the packet network to provide higher data speeds of 115kb/s -384kb/s.

4. *Third Generation* of cellular phone /3G/ allows significantly increased speeds/rates of transmission for data services. 3G phones can be used for Internet access, downloading music & videos, e-mail services and rapid transmitting images from camera phones.

An attempt to establish an international standard for 3G mobile was moderated through ITU, under the auspices of its International Mobile Telecommunications program in 2000 year called IMT-2000, which determined that 3G technology should be based on CDMA systems.

At the ITU's World Radiocommunication Conference in 2000 year, spectrum bands for IMT-2000 systems were allocated on a worldwide basis between 0.8 - 2.5 GHz. Data transfer speeds are in the range 384 kb/s - 2 Mb/s.

By year 2002, the only 3G system in operation was in Japan, although numerous companies have plans for 3G systems in the next years.

The growth in use of mobile telephones has been spectacular. From almost a zero base in the early 1980s, mobile penetration worldwide in 2002 year is estimated at 15.57 mobile phones per 100 people worldwide. But the level of penetration differs between countries. In the United States of America were 44.2 mobile telephones per 100 inhabitants; the penetration rates were 60.53 in France, 68.29 in Germany, 77.84 in Finland and 78.28 in the United Kingdom; the penetration rate in Australia was 57.75, in New Zealand 62.13 and 58.76 in Japan; India had only 0.63 mobile telephones per 100 inhabitants, Kenya had 1.60, China 11.17 and Malaysia 29.95. The number of mobile phones exceeded the number of fixed-wire telephone lines in a variety of countries including Germany, France, United Kingdom, Greece, Italy and Belgium.

However the fixed-lines outnumbered mobiles in the United States of America, Canada and

Argentina. Penetration rates were close to equal in Japan already in 2001 year. But mobile penetration is rising much faster than fixed lines in all countries.

5. *Fourth Generation* of cellular phone /4G/ stands for fourth generation of cellular phone communications standards. In 2008 year, ITU Radiocommunication Sector ITU-R specified the IMT-Advanced requirements for 4G systems to transmit and receive larger amounts of data and at higher rate/speed.

New requirements include increasing throughputs and bandwidths, increased spectrum efficiency and network capacity, lower delays and round-trip times.

The main goal of 4G technology, an all IP based network system, is to provide high speed, high quality, high capacity, security and low cost services for voice and data, multimedia and Internet.

4G aims to support current and emergent multimedia services, social networks and gaming, mobile TV, high-definition television HDTV and video teleconference, digital video broadcast DVB, multimedia messaging service MMS with improved quality of service.

4G wireless communication systems aim to allow indoor and outdoor peak data rates in the range of 1 Gb/s for stationary access and 100 Mb/s for vehicular mobility.

For 4G, North America has used electromagnetic waves bands in the range 700-2600 MHz, South America 2500 MHz, Europe 700-2600 MHz, Asia 800-2600 MHz, Australia & N Zealand 1800-2300 MHz and Multiplexing/Access Technologies: OFDM (orthogonal), MC-CDMA (multiple carrier), LAS-CDMA (large area synchronized), Network-LMDS (local multipoint distribution service).

From 1983 to 2014, worldwide mobile phone subscriptions grew to over seven billion, penetrating 100% of the global population. The wireless networking has become commonplace.

6. *Fifth Generation* of cellular phone /5G/ was initiated in 2015 year because the 4G devices created and consumed growing amounts of data and was designed to process that but also to allow a future wireless standard to address new applications underserved by existed wireless standards, each requiring a new and different set of key performance indicators KPIs. Unlike previous generations of cellular, 5G is more than a new air interface, it is a completely new network architecture.

With mobile traffic expected to increase by a factor of 1,000 over the next decade, 5G technology is the next generation of mobile technology that is due to start arriving in 2020 year.

But 5G is not just another consumer-oriented service providing a lower latency, speedier version of 4G. Amongst its many objectives is an ability to deliver higher capacities and thus enable operators to serve more user devices in a given area. Many of these will be Internet of Things IoT or machine-to-machine M2M services of various kinds. Some will require mains power to drive more demanding M2M applications. Some will be low speed long life services, battery driven. It will offer ultra-low latency for performance-wise applications like gaming, remote brain surgery and updating autonomous cars.

ITU-R coined in 2012 year the International Mobile Telecommunication system for 2020 and beyond called IMT-2020 that defined three cases for high-level use of 5G: enhanced mobile broadband eMBB, massive machine type communication, ultra-reliable machine type communication

eMBB envisions a peak-data rates of 10 Gb/s, 100x over 4G.

In September 2015, ITU-R finalized its vision of the 5G mobile broadband connected society, driving to the standardisation of 5G.

Data rates are empirically linked to available spectrum according to the Shannon-Hartley Theorem, which states that the capacity of transmission channel is a function of bandwidth and channel noise. As solution researchers have looked to electromagnetic waves of higher frequencies offering wider bandwidths.

With the spectrum below 6 GHz fully allocated, the spectrum above 6 GHz, specifically in the mmWave range, presents an attractive alternative to address eMBB use case.

mmWave technology is one of the new 5G technologies and is deployed quickly.

The large amount of contiguous bandwidth available above 24 GHz is needed to meet data throughput requirements and researchers have already shown through prototyping that mmWave technology can be used to deliver data rates above 14 Gb/s.

As the next step in the continuous innovation and evolution of the mobile industry, the fifth generation of cellular phone 5G will not only be about a new air interface with faster speeds, but it will also address network congestion, energy efficiency, cost, reliability and connection to billions of people and devices.

5G brings forward a real wireless world - the Wireless World Wide Web WWWW.

Qualcomm, a company of engineers, scientists and business strategists, envisioned 5G as much more than just a new generation of mobile, but rather a new kind of network that will transform industries, impact economies and societies and ultimately change the world. And to better understand this impact, the company recently commissioned the landmark research project called "The 5G Economy". Also it is collaborating with industry leaders towards 5G commercialization.

Qualcomm vision for 5G is: enhanced mobile broadband, mission critical services, massive Internet of Things IoT

5G will advance mobile from largely a set of technologies connecting people-to-people and people-to-information to a unified connectivity - connecting people to everything.

In 2035 year, when 5G's full economic benefit should be realized across the globe, a broad range of industries - from retail to education, transportation to entertainment and everything between - could produce up to $12.3 trillion worth of goods and services.

The 5G value-chain alone could generate up to $3.5 trillion in revenue and support up to 22 million jobs.

Over time, the total contribution of 5G to the total global **gross domestic product** GDP growth is expected to be the same as that of a country like India, the seventh largest economy in the world. GDP is the monetary measure of the market value of all final goods and services produced in a quarterly/yearly period.

5G dominated the debate at the Mobile World Congress MWC held in Barcelona Spain between 27 February-2 March 2017 from standards development to spectrum harmonisation, from demonstrations to alliances. The year 2017 is the year providing us all an opportunity to think of new ways to manage spectrum beyond the simple split of shared or owned.

Multiple-Input Multiple-Output MIMO is a wireless technology that uses multiple transmitters and receivers to transfer more data at the same time at higher speeds. Massive MIMO will manage interference spatially and share spectrum in both frequency and space like never before. New innovative technologies bring broader bandwidth, faster data rates and longer battery life. A new module, a square device 2.8 inches each side, designed for use in 5G base stations was developed by the companies Ericsson and IBM. It consists of 4 monolithic integrated circuits and 64 dual-polarized antennas.

The biggest benefit of 5G will be enabling mobile operators to deliver data more cost effectively. Despite 5G standards not being expected until 2020 year, twenty-five operators are testing across a wide range of bandwidths, ranging from sub-3GHz to 86GHz. Of the operators that have disclosed their test spectrum, currently the most commonly trialled wavelength is 28GHz, with eight operators using it, as well as 15GHz, which is being used in trials by seven operators. According to Viavi, five operators have reached data rates over 35 Gb/s.

A survey commissioned by the Telecommunications Industry Association TIA and developed in collaboration with subject matter experts from InterDigital and Tolaga Research companies reached the conclusion:

"5G needs to be a *chameleon technology* if aims to achieve a unified network architecture, optimized to support agile business models and diverse applications with wireless connectivity".

TIA is accredited by the American National Standards Institute ANSI to develop voluntary, consensus-based industry standards for a wide variety of Information and Communication Technologies ICT products, and currently represents nearly 400 companies.

7. *Sixth Generation* of cellular phone /6G/ will integrate 5G with telecommunication satellite networks, earth imaging satellite networks and navigation satellite networks, providing multimedia and Internet connectivity, weather information, position identifier, cheap and fast up to 11 Gb/s.

To broadcast such high-speed electromagnetic signals, Nano-Antennas will be implemented at different geographical positions along roadsides, villages, malls, airports, hospitals etc.

Fly Sensors will provide information to remote observer stations, which will check any activity in some areas.

The point-to-point wireless communication networks that transmit super-fast broadband signals through the air will be assisted by high-speed optical fibres lines to broadcast much secured information from transmitters to destinations.

Beside home automation and smart homes/cities/villages with the next step a connected smart globe, the networks would be involved in capturing energy from galactic world, natural calamities control, mind to mind communication.

8. *Seventh Generation* of cellular phone /7G/ would deal with space roaming. A wireless world will demand access to information anytime, anywhere, at high speed with increased bandwidth at minimal cost and with better quality.

references

NIKOLA TESLA - ELECTRICAL WORLD 1917

HOW TO BECOME A WIRELESS OPERATOR - CHARLES B. HAYWARD 1919

RADIO TELEGRAPHY AND TELEPHONY - RUDOLF L. DUNCAN, CHARLES E. DREW 1929

EARTH, RADIO AND THE STARS - HARLAN TRUE STETSON 1934

ULTRA-HIGH-FREQUENCY TECHNIQUES - GLENN KOCHLER, HERBERT J. REICH,
L. F. WOODRUFF & EDITOR J. G. BRAINERD 1942

SHORT WAVE WIRELESS COMMUNICATION - A.W. LADNER, C.R. STONER 1946

INSTRUMENT MANUAL - PAUL H. HUNTER 1947

THEORY AND APPLICATION OF MICROWAVES - ARTHUR B. BRONWELL, ROBERT E. BEAM
1947

FIELDS AND WAVES IN MODERN RADIO - SIMON RAMO, JOHN R. WHINNERY 1953

MICROWAVE ENGINEERING - A. F. HARVEY 1963

BAZELE ELECTROTEHNICII - A. TIMOTIN, A. TUGULEA 1964

FIELDS AND WAVES IN COMMUNICATION ELECTRONICS BY SIMON RAMO,
JOHN R. WHINNERY, THEODORE VAN DUZER 1965

VLF RADIO ENGINEERING - ARTHUR D. WATT 1967

WAVE TRANSMISSION - F.R. CONNOR 1972

TUNNING IN TO NATURE - PHILIP S. CALLAHAN 1977

DIGITAL MICROWAVE RECEIVERS - JAMES BAO-YEN TSUI 1989

TESLA: MAN OF MYSTERY - MICHAEL X 1992

WIRELESS COMMUNICATION - ANDRES LLANA JR 1994

MOBILE RADIO TECHNOLOGY - GORDON WHITE 1994

NAVAL SHIPBOARD COMMUNICATIONS SYSTEMS - JOHN C. KIM, EUGEN I. MUEHLDORF
1995

WIZARD: LIFE AND TIMES OF NICOLA TESLA - MARC SEIFER 1996

WIRELESS PERSONAL COMMUNICATION SYSTEMS - STEPHEN NELSON, DEREK ROGERS,
REG COUTTS 1996

THE PHYSICS OF INFORMATION TECHNOLOGY - NEIL GERSHENFELD 2000

WIRELESS COMMUNICATIONS PRINCIPLES AND PRACTICE - THEODORE S. RAPPAPORT 2002

WIRELESS COMMUNICATIONS & NETWORKS - WILLIAM STALLINGS 2005

WIRELESS COMMUNICATIONS - ANDREAS F. MOLISCH 2005

WIRELESS COMMUNICATIONS - ANDREA GOLDSMITH 2005

MOBILE AND WIRELESS COMMUNICATIONS - GORDON A. GOW, RICHARD K. SMITH 2006

HOW WIMAX WORKS - MARSHALL BRAIN, ED GRABIANOWSKI 2006

TESLA INVENTED RADIO, NOT MARCONI! - LOUIS E. FRENZEL 2007

WIRELESS COMMUNICATIONS FUTURE - WILLIAM WEBB 2007

A GUIDE TO THE WIRELESS ENGINEERING BODY OF KNOWLEDGE (WEBOK) - G. GIANNATTASIO, J ERFANIAN, P.WILLS, H.NGUYEN, T. CRODA, K. RAUSCHER, X. FERNANDO, N. PAVLIDOU, K. D. WONG 2009

ADVANCED POWER MANAGEMENT TECHNIQUES IN NEXT GENERATION WIRELESS NETWORKS - RONNY YONGHO KIM, SHANTIDEV MOHANTY 2010

ENERGY HARVESTING ACTIVE NETWORKED TAGS (EnHANTs) FOR UBIQUITOUS OBJECT NETWORKING - M. GARLATOVA, P.KINGT, I. KYMISSIS, D. RUBENSTEIN, W XIAODONG, G. ZUSSMAN 2010

FUNDAMENTALS OF WIRELESS COMMUNICATION ENGINEERING TECHNOLOGIES - K. DANIEL WONG 2011

THE INCREDIBLE GENIUS THAT AMERICA IGNORED - MICHAEL MICHALKO 2012

A GUIDE TO THE WIRELESS ENGINEERING BODY OF KNOWLEDGE (WEBOK) - EDITOR ANDRZEJ JAJSZCZYK 2012

INCENTIVIZING TIME-SHIFTING OF DATA - S. SEN, C. JOE-WONG, SANGTAE HA, MUNG CHLANG 2012

EVOLVED MULTIMEDIA BROADCAST/MULTICAST SERVICE (EMBMS) IN LTE ADVANCED - DAVID LECOMPTE, FREDERIC GABIN 2012

OVERCOMING SPECTRUM SCARCITY - KATHY PRETZ 2012

LTE DEPLOYMENT: GETTING IT RIGHT THE FIRST TIME - ALLIE WINTER 2012

VINT CERF: A BRIEF HISTORY OF PACKETS - CHARLES SEVERANCE 2012

WIRELESS COMMUNICATIONS - JOSHUA S. GANS, STEPHEN P. KING, JULIAN WRIGHT 2012

TESLA: INVENTOR OF THE ELECTRICAL AGE - W BERNARD CARLSON 2013

TRANSMISSION TECHNIQUES FOR 4G SYSTEMS - MARIO MARQUES DA SILVA, AMERICO CORREIA, RUI DINIS, NUNO SOUTO, JOAO CARLOS SILVA 2013

REGULATIONS CONTINUE TO SHAPE WIRELESS INDUSTRY - DAN MEYER 2013

SMALL CELLS SET TO BE A BIG PART OF LTE - DAN MEYER 2013

WHERE ARE YOU, VoLTE? - ALLIE WINTER 2013

DEVICE-TO-DEVICE COMMUNICATIONS UNDERLYING CELLULAR NETWORKS - D. FENG, LU LU, YI YUAN- WU, G. FENG, S. LI 2013

SCALING MOBILE NETWORK SECURITY FOR LTE: A MULTI-LAYER APPROACH - PATRICK DONEGAN 2014

REVOLUTIONIZING MOBILE ASSURANCE IN THE ERA OF 4G/LTE - PATRICK KELLY, TARA VAN UNEN 2014

IT'S A MOBILE DEVICE REVOLUTION - ELENA NEIRA 2014

TAMING THE COMPLEXITY OF MM-WAVE MASSIVE MIMO SYSTEMS: EFFICIENT CHANNEL ESTIMATION AND BEAMFORMING - STEFANO MONTAGNER, STEFANO TOMASIN 2015

INVESTING IN HETNETS: CARRIER, VENDOR AND ANALYST PERSPECTIVES - MARTHA DEGRASSE 2015

FARADAY AND THE ELECTROMAGNETIC THEORY OF LIGHT - AUGUSTO BELENDEZ 2015

WHAT TECHNOLOGY INVENTIONS ARE REQUIRED TO MAKE 5G NEW RADIO NR A REALITY - JOHN SMEE, MATT BRANDO 2016

mmWAVE: BATTLE OF THE BANDS - SARAH YOST NI 2016

LiFi UNLOCKING UNPRECEDENTED WIRELESS PATHWAYS FOR OUR DIGITAL FUTURE - HARALD HAAS, NIKOLA SERAFIMOVSKI 2016

EVOLUTION OF MOBILE COMMUNICATION FROM 1(G) TO 4G, 5G, 6G, 7G ... - AARTI DAHIYA 2016

THE ESSENTIAL ROLE OF GIGABIT LTE & LTE ADVANCED PRO IN A 5G WORLD - BASMUS HELLBERG, SUNIL PATIL 2017

TEN COMMUNICATION TECHNOLOGY TRENDS FOR 2017 - ALAN GATHERER 2017

PREPARING LINEARITY AND EFFICIENCY FOR 5G - NOEL KELLY, WENHUI CAO, ANDING ZHU 2017

5G FUTURE - STEVE MOLLENKOPF , QUALCOMM 2017

ERICSSON AND IBM ANNOUNCE 5G BASE STATION CHIP - MARTHA DEGRASSE 2017

TIGHT FOCUS TOWARD THE FUTURE - NILS WEIMANN, MARUF HOSSAIN, VIKTOR KROZER, WOLFGANG HEINRICH, MARCO LISKER, ANDREAS MAI, BERND TILLACK 2017

PUSHING THE ENVELOPE FOR HETEROGENEITY - KAMAL K. SAMANTA 2017

HOW LOW POWER WIDE AREA NETWORKS LPWANs ARE REVOLUTIONISING THE WIRELESS WORLD - BOYD MURRAY 2017

25 MOBILE OPERATORS ALREADY TESTING 5G TECHNOLOGY - GUY DANIELS 2017

mmWAVE AND MASSIVE MIMO IN NEXT-GENERATION WIRELESS SYSTEMS - MODERATOR: MIGUEL DAJER 2017

IEEE RELEASES DETAILS ABOUT ITS 5G AND BEYOND ROADMAP - KATHY PRETZ 2018

LIVE SCIENCE

WEBOPEDIA

TECHNOPEDIA

WIKIPEDIA

WIKIMEDIA COMMONS

PREVENTING MALADIES
PROTECTING HEALTH

An early object with a history dating back 9000 years in the country of Mesopotamia is the pillow made from stone blocks, designed more to keep bugs out of the ears than for comfort. The pillows changed over time in shapes, materials, purpose and meaning presenting an interesting evolution from more viewpoints.

Ancient Chinese pillows were more decorative and made from bamboo, porcelain and bronze. Romans and Greeks of Ancient Europe used pillows made from cotton filled with cotton, reeds or straw, to support the head in bed and to have some comfort.

In the Middle-Ages the pillows were made from linen and cotton filled with straw or goose dawn, but were not often used because they were considered a sign of weakness.

The first Industrial Revolution about 1800 year, facilitated the mass production of affordable pillows and so they became used in every household and over time, the history of the pillow has seen many different pillow shapes and sizes, whilst serving many alternative uses.

Today, by undertaking a revolutionary process, the pillows are made from a variety of different fabrics such as cotton, linen, polyester and beside down or feather use superior fillings to offer a high level of support and comfort. There is a hypo-allergic option for people who are allergic to usual down and feather fillings. There are neck pillows known as orthopaedic or cervical pillows, which offer additional neck and head support. There are also body pillows from 3 to 5 feet in length to offer support to the neck, legs, knees and abdominal area.

Any pillow you chose to sleep with each night should offer you an optimal comfort level so you can wake feeling well rested and refreshed every time.

And if the pillow can prevent you to become ill, it would be the ultimate achievement.

The novelty pillows presented further have been specially devised in order to provide new types of pillow of simple, cheap, effective and readily construction and whereby the head is comfortable rested, allowing to be protected from cold air and maintained at body temperature, very suitable in cold nights, cold weather, cold regions and preventing maladies like cold, flu, migraine, sinusitis, otitis, meningitis, encephalitis, tumour etc. while it is allowing to breath comfortably. They are: "Pillow Variant" with *IPA innovation patent nr. 2010101399 and "Hood Pillow" with IPA innovation patent nr. 2011101637

Their serial production entitles the author to royalties.

*IPA = Intellectual Property Australia

PILLOW VARIANT

The following statement is a full description of this invention, including the best method of performing it known to me:

It is very known that a pillow is a rectangular cushion, which supports the head when you are resting in bed. This invention has been specially devised in order to provide a variant for pillow of simple, cheap, effective and readily construction and whereby the head is protected from cold air and may be warmly kept and any unintentional contact with cold air is prevented.

A *pillow variant* in accordance with this invention comprises an ordinary pillow with another ordinary pillow mounted over it to up edge by sewing them at the three matching edges.

The opening between the two pillows is of such a size and in such a position that when the head is resting on the first pillow, its upper part is covered by the second pillow, together mounted pillows are an assembly acting like a hood allowing the head to be protected from cold air and maintained at body temperature, very suitable in cold nights, cold weather, cold regions and preventing illnesses like cold, flu, migraine, sinusitis, otitis, meningitis, encephalitis, tumour etc. while it is allowing to breath comfortably.

The first pillow is of width W and height H and the other pillow is of width W and height H/2.

CLAIMS

The claims defining the invention are as follows:

1. A *pillow variant* comprising an ordinary pillow size WxH with another ordinary pillow size WxH/2 mounted over it to up edge by sewing them at the three matching edges.
2. A *pillow variant* as claimed in claim 1 is an assembly with an opening so that when the head is resting on the first pillow, its upper part is covered by the second pillow.
3. A *pillow variant* according to claims 1, 2 wherein the assembly is made from cotton and filled with down. Other suitable materials can be used.
4. A *pillow variant* substantially or herein before described with reference to Figures 1-3 of the accompanying drawings.
5. A *pillow variant* according to claims 1, 2 wherein the assembly prevents the direct contact with cold air in the time of resting while allowing a comfortable breath.

ABSTRACT

The disclosed *pillow variant* is the assemblage of an ordinary pillow 1 size WxH with another pillow 2 size WxH/2 mounted over it to up edge by sewing the matching edges 3, those pillows forming an assembly with an opening 4 of such a size and in such a position that when the head is resting on the first pillow, its upper part is covered by the second pillow, together mounted pillows acting like a hood preventing the direct contact with cold air in the time of resting with a comfortable breath.

DRAWINGS

The invention may be better understood with reference to the illustrations of embodiments of the invention which:

Figure 1 is a view of an ordinary pillow of width W and height H.

Figure 2 is a view of another ordinary pillow of width W and height H/2.

Figure 3 is a view of a pillow variant, the assemblage of the pillows presented in Figures 1 and 2 in conformity with the invention.

The *pillow variant* is the assemblage of an ordinary pillow 1 of size W x H with another pillow 2 of size W x H/2 mounted over it to up edge by sewing the matching edges 3, those pillows forming an assembly with an opening 4 of such a size and in such a position that when the head is resting on the first pillow, its upper part is covered by the second pillow, together mounted pillows acting like a hood preventing the direct contact with cold air in the time of resting while allowing a comfortable breath

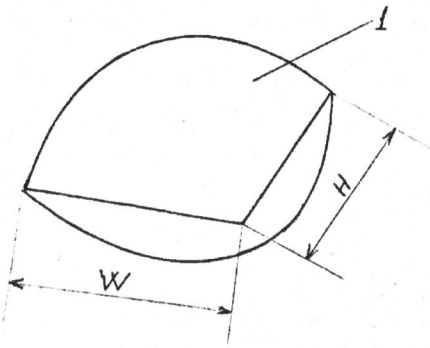

FIG.1

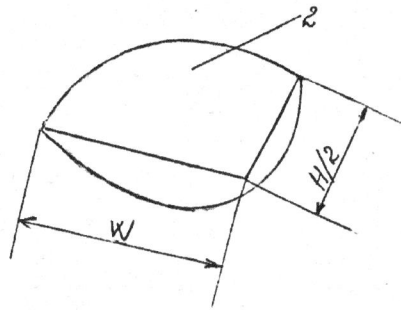

FIG.2

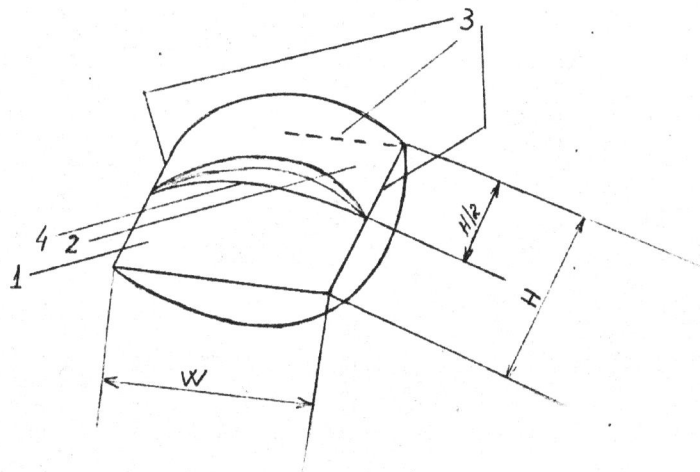

FIG.3

71

HOOD PILLOW

The following statement is a full description of this invention, including the best method of performing it known to me:

It is very known that a pillow is a rectangular cushion that supports the head when you are resting in bed. This invention has been specially devised in order to provide a variant for pillow of simple, cheap, effective and readily construction and whereby the head is protected from cold air and may be warmly kept and any unintentional contact with cold air is prevented while allowing a normal breath.

A *hood pillow* in accordance with this invention comprises an ordinary pillow of size W x L with open mounted zips, one on the lateral left edge and one on lateral right edge, on 2L/3 from the top allowing by zipping up to transform the pillow in a hood.

The opening of the *hood pillow* is of such a size and in such a position that when the head is resting its upper part is covered by the upper part of the pillow, the assembly acting like a hood allowing the head to be protected from cold air and maintained at body temperature, very suitable in cold nights, cold weather, cold regions and preventing illnesses like cold, flu, migraine, sinusitis, otitis, meningitis, encephalitis, tumour etc. while allowing to breath comfortably.

The ordinary pillow is of width W and of length L and the zips are mounted on the lateral edges on 2/3 of length from top; by zipping up, resulting device *hood pillow* is of width W and of length 2L/3 (size W x 2L/3) with an opening for the upper part of the head of size W x L/3.

The hood pillow is made from cotton and filled with down. Other suitable materials can also be used. A hood pillow can also be in accordance with this invention by replacing the zips in fastened position by sewing.

CLAIMS

The claims defining the invention are as follows:

1. The *hood pillow* in accordance with this invention comprises an ordinary pillow with two zips, one on the left edge and one on the right edge, opened mounted on 2/3 of the edges from the top allowing by zipping up to transform the pillow in a hood.
2. The *hood pillow* as claimed in claim 1 is an assembly with an opening so that when the head is resting on the pillow, its upper part is covered by the upper part of the pillow. It protects reliably against cold air while permits a normal breathing.
3. The *hood pillow* according to claims 1, 2 wherein the assembly is made from cotton and filled with down. Other suitable materials can be used.
4. The *hood pillow* substantially or herein before described with reference to Figures 1-2 of the accompanying drawings.
5. The *hood pillow* as claimed in claim 1 where the zips in fastened position can be replaced by sewing.

ABSTRACT

The *hood pillow* in accordance with this invention comprises an ordinary pillow of suitable size with zips on the lateral left and right edges, open mounted on 2/3 of the edge length from the top, allowing by zipping up to transform the pillow in a hood.

The disclosed *hood pillow* is the resulting assembly of an ordinary pillow 1 with zips on the lateral edges mounted opened on 2/3 of their length from the top, allowing by zipping up to transform the pillow in a device with an opening 2 of such a size and in such a position that when the head is resting on the pillow, its upper part is covered by the upper part of the pillow like in a hood preventing the direct contact with cold air in the time of resting while allowing a normal breathing.

DRAWINGS

The invention may be better understood with reference to the illustrations of embodiments of the invention which:

Figure 1 is a view of an ordinary pillow of width W and length L with complete opened zips on 2L/3 length on each of lateral edges.

Figure 2 is a view of the resulting assembly in conformity with the invention, the *hood pillow* of width W and length 2L/3.

The *hood pillow* is the device resulting from the ordinary pillow 1 by zipping up 2/3 length of each lateral edge left and right (by themselves), the assembly having an opening 2 of such a size and in such a position that when the head is resting, its upper part is covered by the upper part of the pillow, the resulting device acting like a hood preventing the direct contact with cold air in the time of resting while allowing a normal breathing.

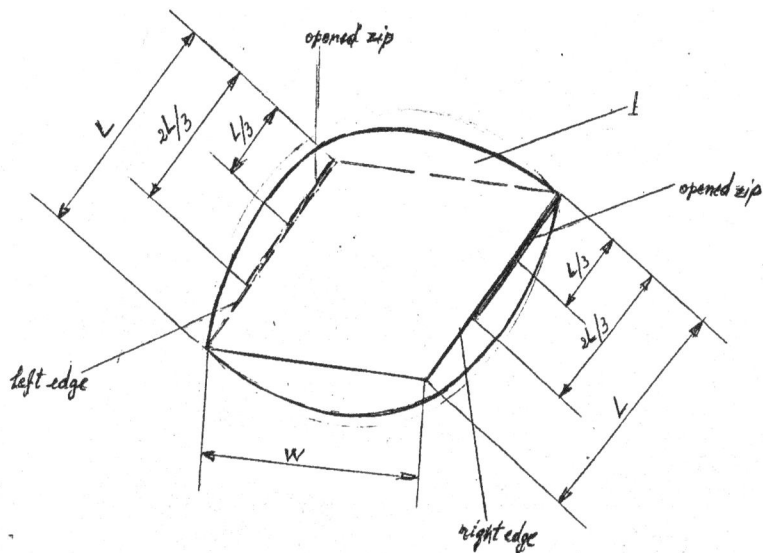

FIG 1

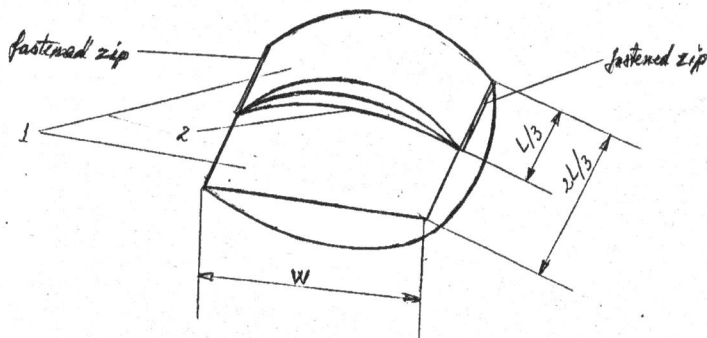

FIG 2

73

SCIENCE AFTER
MARIE CURIE'S DISCOVERIES

When the French physicist **Henri Becquerel** (1852-1908) discovered the uranium rays in the year 1896 and when the Polish-French scientist **Marie Sklodowska Curie** (1867-1934) began to study them in 1897 year, there was no knowledge about the origin of X-rays or uranium rays and in the physical science was given that the atom was indivisible and unchangeable. But the work of Marie Sklodowska Curie led scientists to suspect that the theory that the atom was indivisible and unchangeable was untenable.

The savante Marie Sklodowska Curie was born in the second part of the 19th century, in 1867 year and she began her research work on uranium rays at the end of 19th century, in 1897 year. The 19th century was a period, which experienced rapid progress in science and technology; important breakthroughs were in the iron & steel technology, electricity, weapons, physics & chemistry, sociology, psychology and biology. In the study of physics, was better understanding of the nature of matter. There was revived the atomistic theory, references to the concept of atom dating back to ancient times or antiquity.

Inspired by the spiritual activities of ancient Greece in pursuit of a novel form of natural philosophy, the Greek philosopher **Leucippus** (~ 440 BC) and his student, **Democritus** (460-370 BC) did one of the most amazing intellectual accomplishments of the antiquity, closer to the truth than anyone else in the following millennium. Leucippus and Democritus held that the nature of things consists of an infinite number of extremely small particles, which they called "atoms", physically indivisible, indestructible and full - containing no empty space inside them - and the "void" that surrounds them.

The Greek word *"ἄτομος"* (atomos) means indivisible, unable to be broken further.

Leucippus and Democritus were atomists and according to the atomists, Nature exists only of two things, the *atoms* and the *void* that surrounds them. Because of their indestructibility, the *atoms* were considered eternal and undergoing a continual and endless random motion, determined by mutual collisions.

The fundamental concepts of the theory seem to have been formulated by Leucippus and historically Leucippus is considered the father of the theory.

The atomic hypothesis, born 25 centuries ago, remained for the next 23 centuries essentially a vision of the imagination. No empirical evidence based on either observation or experimentation existed to prove or disprove the key hypothesis of the theory – the corpuscular structure of matter. It was first a philosophical idea, its philosophical epoch is ancient, but its scientific epoch is relatively recent.

Only in the 17th century, appears something similar to the theory of atomism, the *corpuscularianism*, a physical theory that supposed that all matter to be composed of minute particles, corpuscles, which could in principle be divided . The Latin word "corpusculum" means small body.

Further progress in the understanding of matter did not occur until the science of chemistry began to develop and was to change suddenly and profoundly at the beginning of the 19th century with the birth and consolidation of scientific atomism in the modern sense of phrase. Its birth or its growth proved not easy, experiencing many vicissitudes in its accumulation of empirical data, which often suffered from the imperfection of the research tools and the difficult maturation of theoretical concepts.

In 1789 year the French nobleman and scientific researcher, father of modern chemistry **Antoine-Laurent de Lavoisier** (1743-1794) discovered the law of conservation of mass and defined an *element* as a basic substance that could not be further broken down by the methods of chemistry.

The English chemist, meteorologist, physicist and schoolteacher **John Dalton** (1766-1844) in 1805 year, revived the atomistic theory using the concept of atoms to explain why elements always react in ratios of small whole numbers (the law of multiple proportions) and why certain gases dissolved better in water than others. He proposed that each element consists of atoms of a single, unique type and that these atoms can join together to form chemical compounds.

John Dalton is considered the originator of modern atomic theory.

Various atoms and molecules are depicted in John Dalton's book "A New System of Chemical Philosophy" appeared in 1808 year, an early scientific work on modern atomic theory, which is considered the birthday of the modern atomic theory. In book, John Dalton proposed that atoms have weight. He postulated the theory in which the atom was conceived as being a tiny billiard ball; the materials are made from *atoms*; a material of the same atom is an *element*; a material of different atoms is a *compound (substance)*.

Dalton theorized that the elements combine in fixed ratios into compounds, as for ex. in water are two atoms of hydrogen always combined with one atom of oxygen.

Element or *chemical element* is called the simplest thing in Nature which stays at the base of all other things that surround us, thought of as basic chemical building block of matter; it is a material which cannot be broken down or changed into another substance using chemical means.

Dalton proceeded to print his first published *table of relative atomic weights* with six elements namely hydrogen, oxygen, nitrogen, carbon, sulphur and phosphorus, with the atom of hydrogen conventionally assumed to weight 1.

By the dawn of the 20th century, the chemists were engaged in experiments with atoms of the known elements and the entity of "atom" seemed almost universally accepted.

It looked like the Greek philosophers Leucippus and Democritus of antiquity, by holding that the nature of things consists of an infinite number of extremely small particles which they called "atoms", physically indivisible, indestructible and full, finally won the battle.

However suddenly the work of Marie Sklodowska Curie led scientists to suspect that the theory that the atom is indivisible and unchangeable is untenable. Marie Sklodowska Curie proved in her experiments that the radiation of uranium compounds can be measured with precision under determined conditions. She proved that the radiation of uranium compounds depends neither on conditions of chemical combination nor on external circumstances such as light or temperature and that this radiation is an atomic property of the uranium element, something coming from the internal structure of its atoms, meaning that the atoms are not indivisible and indestructible.

The atoms had lost their traditional and fundamental characteristic of indivisibility. Even though they retained their role as distinct and individual constituents of all observable substances, atoms could not longer be considered the most fundamental particles of matter.

The victory of the classical atomic theory proved short-lived.

Was found also that the radiation of the chemical element thorium has an intensity of the same order as that of uranium and is, as in the case of uranium, an atomic property of the element.

At this point, it was necessary to find a new term to define this new property of matter manifested by the elements uranium and thorium. Marie Curie proposed the word *radioactivity,* which has since become generally adopted; the radioactive elements have been called *radio elements.*

During the course of her research, Marie Curie examined not only simple compounds, salts and oxides, but also a great number of ores. Certain ones proved radioactive; these were those containing the elements uranium and thorium called uraninite/pitchblende; but their radioactivity seemed abnormal since was much greater than the amount found in uranium and thorium.

Marie Curie did then the hypothesis that the ores containing uranium and thorium contain too in small quantity a substance much more strongly radioactive than either uranium or thorium.

Soon became recognized experimentally the presence in the pitchblende of two new radioactive chemical elements: polonium and radium.

A new chapter in the history of atomic theory began with the discovery that the atoms, these individual building blocks of matter, are themselves made of a number of smaller components. While they still remained the basic units of the chemical elements, atoms turned out to be themselves composite entities.

The work of Marie Sklodowska Curie opened the door to yet unknown intrinsic world of atom and strongly influenced the science in the years that followed.

Consequently the atom of radium would be in a process of evolution and we should be forced to abandon the theory of the invariability of atoms, which is at the foundation of modern chemistry. Moreover, we have seen that radium acts as though it shot out into space a shower of projectiles, some of which have the dimensions of atoms, while others can only be very small fractions of atoms. If this image corresponds to a reality, it follows necessarily that the atom of radium breaks up into subatoms of different sizes, unless these projectiles come from the atoms of the surrounding gas, disintegrated by the action of radium; but this view would likewise lead us to believe that the stability of atoms is not absolute.

(from article Radium and Radioactivity by Marie Curie, 1904)

MARIE SKLODOWSKA CURIE
1903 PHYSICS NOBEL PRIZE RECIPIENT
1911 CHEMISTRY NOBEL PRIZE RECIPIENT

Marie Sklodowska Curie's findings put her in the front row of the world scientists and stimulated them to explore, investigate, research for new elements and for explanations of unknown phenomena. The scientists have imagined, theorised and experimented new ways of understanding the matter and found areas of common inferences drawn from collected data, leading to modern explanations of nature that could direct to new others anytime.

In the years 1899-1900, the New Zealander chemist and physicist **Ernest Rutherford, First Baron Rutherford of Nelson** (1871-1937) working at McGill University in Montreal, Canada and the French chemist and physicist **Paul Villard** (1860-1934) working in Paris, France separated the *radioactive radiation* of the elements uranium and thorium into three types, eventually named by Ernest Rutherford *alpha α, beta β* and *gamma γ*.

The separation was based on the penetration of objects and the deflection by a magnetic field and was demonstrated that some of the mysterious rays emanating from radioactive substances were not rays at all, but tiny particles. The scientists showed that the radioactive atoms emit three different kinds of radiation:

1. One kind of radiation, called *alpha radiation,* is particles of matter.

An alpha particle has a positive electric charge and its mass is about four times the mass of a hydrogen atom. Alpha particles exit radioactive atoms with high energies, but they lose this energy as they move through matter. An alpha particle passes through a thin sheet of aluminium foil, but is stopped by anything thicker.

2. A second kind of radiation, called *beta radiation*, turned out to be very light particles with a negative electric charge. The beta particles travel at nearly the speed of light and can make their way through half a centimetre of aluminium.

3. A third kind of radiation, called *gamma radiation,* is true rays, same kind of thing as radio waves and light, with no mass and no electrical charge. They are similar to the X-rays but more energetic, an energetic form of electromagnetic radiation. Gamma rays emitted by radioactive atoms can penetrate deeper into matter than alpha or beta particles. A small fraction of gamma rays can pass through a meter of concrete.

So was proved that radioactivity is emission of tiny particles and energetic waves from atom.

Building on Marie Sklodowska-Curie research, the scientists soon realized that if atoms emitted such things they couldn't be indivisible or unchangeable: atoms are made up of smaller particles. In that time any laboratory where people worked with radium or other radioactive minerals, had radioactivity tending to spread around everywhere; the labs were contaminated by an"emanation" of the radioactive compounds, a radioactive gas, which also gave rise to the radioactivity of solid substances on which it fell, named "excited radioactivity". In 1900 year, Ernest Rutherford found that the radioactivity of the *emanation* from thorium diminishes or decays with time.

Working with the English radiochemist **Frederick Soddy** (1877-1956) at McGill University in Canada, in 1902-1903 years Rutherford identified the phenomenon of *radioactive half-life*. The radioactivity of gases of radioactive compounds, radioactivity of emanations, disappears spontaneously according to an exponential law with a time constant, which is characteristic for each radioactive element: the emanation from radium diminishes by one-half every 4 days, the emanation from thorium diminishes by one-half every 55 seconds and the emanation from actinium diminishes by one-half every 3 seconds.

The decay of radioactivity was a vital clue for the physicists Ernest Rutherford and Frederick Soddy to develop a *revolutionary hypothesis* to explain the process. Together they realized that *radioactive elements can spontaneously change into other elements. As they do so, they emit radiation of one type or another. The spontaneous decay process continues in a chain of emissions until a stable atom is formed.*

The two physicists realized that, that was the *transmutation of elements* that had eluded alchemists for thousands years. They realized that the ceaseless radiations come from a *vast store of energy within atoms*.

In the Rutherford's picture of transmutation, a radium atom emits an alpha particle, turning into "emanation" (in fact the gas radon), this atom in turn emits an alpha particle to become "radium A"(now known to be a form of polonium) and the chain eventually ends with stable lead.

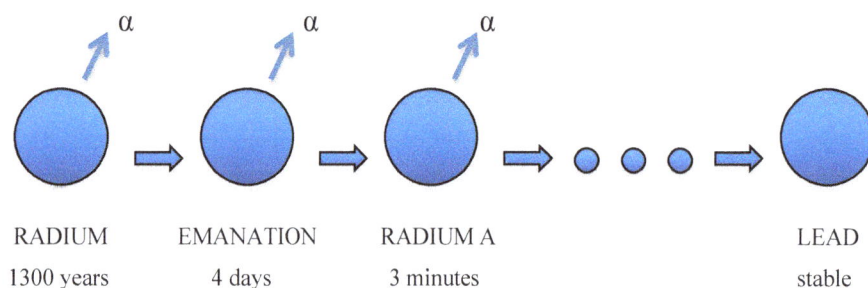

RADIUM	EMANATION	RADIUM A		LEAD
1300 years	4 days	3 minutes		stable

This theory stimulated the scientists to order all known radioactive elements into their decay series and to search for any missing members.

Ernest Rutherford is known as the father of nuclear physics and considered the greatest experimentalist since Michael Faraday. He was awarded **1908 Nobel Prize in Chemistry** for his investigations into the disintegration of the elements and the chemistry of radioactive substances.

ERNEST RUTHERFORD
1908 CHEMISTRY NOBEL PRIZE RECIPIENT

In 1905 year, Ernest Rutherford would marvel about the unexpected then advance of science: "The rapidity of this advance has seldom, if ever, been equalled in the history of science. By the end of this century physics would transform nearly every aspect of everyday life. It would also alter our vision of reality, the universe we inhabit and our peculiar and enigmatic place on it".

The light, especially the ultra-violet light, discharges negatively electrified bodies with the production of rays of the same nature as cathode rays. Under certain circumstances the light can directly ionize gases. The German physicists **Heinrich Rudolf Hertz** (1857-1894) and **Wilhelm Ludwig Franz Hallwachs** (1857-1922) discovered the first of these phenomena in 1887 year, while the second of these phenomena was announced first by the German physicist **Philipp Eduard Anton von Lenard** (1862-1947) in 1900 year.

In 1902 year, Philipp Lenard observed that the outcome increased with the frequency of the light. Classical physics laws could not explain these phenomena, known as "Hertz effect or photoelectric effect".

From 1888 year, Lenard contributed a lot to the cathode ray theory by his research on cathode rays, which he called "quanta of electricity" or shortly "quanta".

Continuing his investigation for the origin of the cathode rays, after concluded in 1897 year that they were composed of very light, negatively charged energetic particles, previously unknown, which he called "corpuscles", the British physicist **Sir Joseph John Thomson** (1856-1940) showed that the corpuscles were identical with particles given off by photoelectric and radioactive materials.

Soon was the general realization that quanta were constituent parts of the atom. It was also soon recognised that the cathode rays are the particles that carry electric currents in metal wires and carry the negative electric charge of the atom.

The Anglo-Irish physicist **George Johnstone Stoney** (1826-1911) postulated in 1874 year and again in 1881 year a "fundamental unit for quantity of electricity", named by him the "atom of electricity". His most important scientific work was the conception and calculation of that's magnitude. In 1891 year, George Johnstone Stoney proposed another term, "electron" with symbol "e" to describe the fundamental unit of electrical charge.

After 1900 year, *quanta* and *corpuscles* became associated with the fundamental unit of electrical charge and eventually named *electrons*.

In the second part of the nineteenth century, 1877 year, the Austrian physicist and philosopher **Ludwig Eduard Boltzmann** (1844-1906) presented the theoretical possibility that the energy states of a physical system could be discrete.

A physical system has a property called *energy* E and a corresponding property called *mass* m. The two properties are equivalent in that they are always both present in the same, constant proportion to one another.

The concept that the mass of an object or system is a measure of its energy content is called *mass–energy equivalence.*

Mass–energy equivalence arose originally as a paradox described by the French mathematician, theoretical physicist, engineer and philosopher of science **Jules Henri Poincaré** (1854-1912).

In 1900 year, Henri Poincaré discovered a relation between mass and electromagnetic energy. Henri Poincaré concluded that the electromagnetic field energy E of an electromagnetic wave behaves like a fictitious fluid "fluide fictif" with a mass density of E/c^2 (c is the speed of light).

Studies on spectrum of thermal emission (heat) from objects, frequently called the blackbody function, led to physics laws. A blackbody is an idealized physical system (body) that absorbs all incident electromagnetic radiation, regardless of frequency or angle of incidence.

In 1900 year, the German theoretical physicist **Max Karl Ernst Ludwig Planck** (1858 –1947) *originated* the *quantum theory* in the famous *Planck black-body radiation law.* The central assumption presented to the Deutsche Physikalische Gesellschaft (DPG) on 14 December 1900 year was the supposition, now known as the *Planck postulate*, that the energy radiated by a black body could be emitted only in quantized form $E_{rad} = nE$ where n is an integer n=1,2,3...

In other words 'the energy could only be a multiple of a "quantum" or an elementary unit $E = hf$ where h is Planck's constant, also known as Planck's "action quantum" and f is the frequency of the radiation.

The Latin word "quantum" means "how much" and its plural is "quanta".

At first Planck considered that his *quantisation* was only "a purely formal assumption".

Nowadays this assumption, incompatible with classical physics, is regarded as the *birth of Quantum Physics* and the greatest intellectual accomplishment of Planck's career.

The discovery of Planck's constant enabled Planck to define a new universal set of physical units, such as the 'Planck length' and the 'Planck mass', all based on fundamental physical constants on which much of quantum theory is based.

Max Planck was awarded the **1918 Physics Nobel Prize** in recognition of his fundamental contribution to a new branch of physics.

The heat and the light are common ways that we experience electromagnetism, the combination of electricity with magnetism, in our daily lives.

Alongside Max Planck's work on *quantum of heat*, the German-American theoretical physicist **Albert Einstein** (1879-1955) in March 1905 year, created the *quantum of light* in a paper that "explained the *photoelectric effect* as result of light energy being carried in discrete/quantized packets".

Albert Einstein also presented the concept that a body losing energy as radiation or heat was losing mass of amount m=E/c^2, solving Poincaré's paradox without using any compensating mechanism.

Equivalence of energy E and mass m for a body is described by the famous equation E=mc^2 that expresses the theory of the equivalence of Matter and Energy for any physical system and serves to convert units of mass to units of energy and vice-versa, no matter what system of measurement units is used.

Albert Einstein was awarded the **1921 Physics Nobel Prize** for the discovery of the law of the photoelectric effect.

Later in 1914 year, the American experimental physicist **Robert Andrews Millikan** (1868-1953) confirmed experimentally Einstein's law on photoelectric effect and measured the electric charge of the electron e = $-$ 1.592 x 10^{-19}C. Today the accepted value is 1.602 x 10^{-19}C. Robert Millikan was awarded the **1923 Physics Nobel Prize** for measuring the electric charge of the electron and for the work on photoelectric effect.

Further experiments, starting with the American physicist **Arthur Holly Compton** (1892-1962) demonstrated the particle concept of the electromagnetic radiation.

Arthur Compton found in 1922 year, that X-ray electromagnetic radiation, after scattered by free electrons had longer wavelength. Compton explained that by assuming a particle nature for X-ray and by applying conservation of energy and conservation of momentum to the collision between the "X-particle" and the electron. The scattered X-ray has lower energy and therefore a longer wavelength, the difference of energy having been transferred to the electrons, discovery known as the "Compton effect" or "Compton scattering". Arthur Holly Compton was awarded the **1927 Physics Nobel Prize** for his scattering studies, which demonstrated the particle nature of the electromagnetic radiation.

In 1926 year, the French optical physicist **Frithiof Wolfers** (1890-1971) and the American chemist **Gilbert Newton Lewis** (1875-1946) coined the name "photon" for the light particles. The name is derived from Greek word "φώτο" (photo) meaning light and the "on" at word end indicates that it is a particle belonging to same class as the electron and the proton. *Photon* is defined as the elementary/fundamental particle of electromagnetism, the *quantum of light and all other forms of electromagnetic radiation*, symbol "γ".

It has zero rest mass and energy and moves at a constant velocity c=2.9979x108 m/s in free space, its energy at a certain frequency f is equal to hf, allows long distance interactions, exhibit wave–particle duality, can be destroyed/created when radiation is absorbed/emitted and can have particle-like interactions with other particles.

The photon belongs to the class of elementary particles that carries the fundamental interactions of nature called "gauge bosons". The photons carry the electromagnetic force. The photons carry the electric forces that hold atoms together, forces between the atom's electric charges.

The photons within the atom are transitory entities existing on timescales too fine to be perceived by our macroscopic senses.

ATOM STRUCTURE MODELS

When Marie Curie discovered radium and published her results to the world, what is called the Atom Age was beginning to dawn. The problem of matter structure started to interest many scientists and many atomic models were developed and studied with zeal. Most of these had more significance for chemistry than for physics but all have paved the way for real physical investigation.

The American physical chemist Gilbert Newton Lewis developed in 1902 year the theory of the *cubical atom,* an early, *first atomic model* in which structure the electrons were positioned at the eight corners of a cube in a non-polar atom or molecule, theory based on "Abegg's rule". In chemistry, Abegg's rule states that the sum of the absolute value of the negative valence and the positive valence of an element is frequently eight. Valence is the amount of power of an atom determined by the number of electrons the atom will lose, gain, or share when forms compounds.

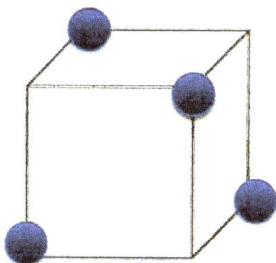

Cubical atomic model for element Carbon by G.N. Lewis (1902)

Above is shown an example of model for the atom of chemical element Carbon, belonging to group 4/row 2 of the periodic table of chemical elements.

Philipp Eduard Anton von Lenard, following latest findings that in electrically neutral atom there are negatively electric charged electrons, postulated in 1903 year that the fundamental building block of all elements is a couplet of a positive and a negative electric charge matter bound together, which he called the *dynamid*. The atomic mass of an atom is proportional to the number of dynamids present, that is a hydrogen atom is a single dynamid, helium would be four dynamids and so on. The dynamids were surrounded by a field of electric force but the binding force, which holds those dynamids together, was not explained. About the arrangement in space of dynamids, Lenard suggested nothing but claimed correctly that for their most part atoms consist of empty space.

Philipp Lenard was awarded **1905 Physics Nobel Prize** for research on cathode rays and their properties. Below it is shown an example for atom structural representation proposed by Philipp Lenard.

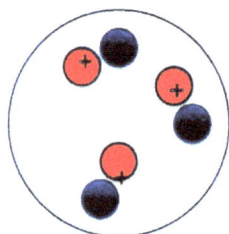

Dynamid atomic model by P. Lenard (1903)

Joseph John Thomson proposed in 1904 year the *plum pudding model* for atom: an atom was a composite in which a number of electrons (which Thomson still called "corpuscles") are imbedded in a wad of undifferentiated positive matter "like raisins in a pudding".

Thomson postulated that the low mass, negatively charged electrons were distributed throughout the atom non-randomly, possibly rotating in rings, with their charge balanced by the presence of a uniform sea of positive electric charge.

In 1905 year Thomson discovered the natural radioactivity of the chemical element potassium and in 1906 year he demonstrated that the element hydrogen had only a single electron per atom, previous theories allowing various numbers of electrons.

Josef John Thomson was awarded the **1906 Physics Nobel Prize**.

Further it is shown an example for atom structural representation proposed by Thomson.

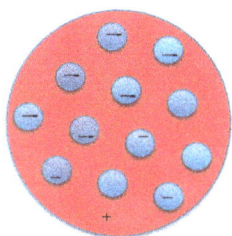

Plum pudding model of atom by J.J. Thomson (1904)

The Japanese physicist **Hantaro Nagaoka** (1865-1950), pioneer of Japanese physics, rejected Thomson's atom model on the ground that opposite electric charges are impenetrable. He proposed an alternative model in which a positively electric charged centre is surrounded by a number of revolving electrons, in the manner of Saturn planet and its rings.

Nagaoka's *Saturnian model* of atom, as shown farther, was developed in 1904 year and was based around an analogy to the explanation of the stability of the Saturn rings - the rings are stable because the planet they orbit is very massive. The model made two predictions: very massive nucleus (in analogy to a very massive planet) & electrons revolving around the nucleus, bound by electrostatic forces (in analogy to the rings revolving Saturn, bound by gravitational forces).

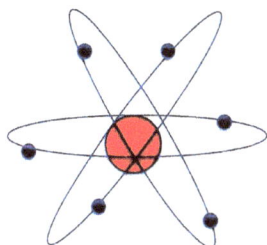

Saturnian model of atom by Hantaro Nagaoka (1904)

In the year 1906, the English physicist **Lord Rayleigh / John William Strutt, third Baron Rayleigh of Terling Place** (1842-1919) considered that an infinite number of electrons in the plum pudding atom model of Thomson could lead to idea that the cloud of electrons be assimilated to a fluid whose properties differ from those of known fluids. He proposed an *electron fluid model* for atom.

Lord Rayleigh and the Scottish chemist **Sir William Ramsay** (1852-1916) investigated the densities of the most important gases and in connection with these studies discovered the chemical element *argon.* Lord Rayleigh was awarded the **1904 Physics Nobel Prize**.

The proposed *electron fluid model* for atom is presented further.

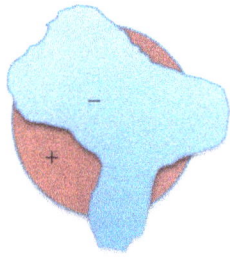

Electron fluid model of atom by Lord Rayleigh (1906)

The English physicist, astronomer and mathematician **Sir James Hopwood Jeans** (1877-1946) proposed in 1906 year the *vibrating electron model* of atom, presented below, by considering the electron itself an object with its own internal structure, a 'structured electron', whose vibrations could lead to radiation of different frequencies emitted by atoms.

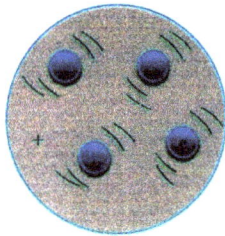

Vibrating electron model of atom by J.H. Jeans (1906)

The British mathematician **George Adolphus Schott** (1868-1937) suggested in 1906 year like Jeans, that the electron must be more complicated than a point particle and the radiation of different frequencies emitted by atom results from the electron expanding at a very slow rate, proposing the *expanding electron model* of atom, presented farther.
Schott's work says nothing about the positive charge in the system.

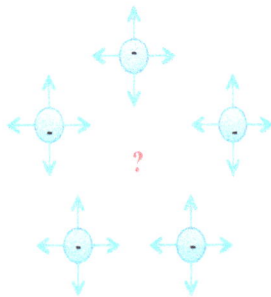

Expanding electron model of atom by G.A. Schott (1906)

The German physicist **Johannes Stark** (1874-1957) proposed later, in 1910 year, the dipole *archion* as fundamental component of atom, acting as a microscopic permanent bar magnet, but also positive charged electrically. The electrons were added somewhere in the vicinity of the archions, to neutralize the repulsion between positive archions and hold the entire structure together. He proposed the *archion model* of atom whose structure is shown below.

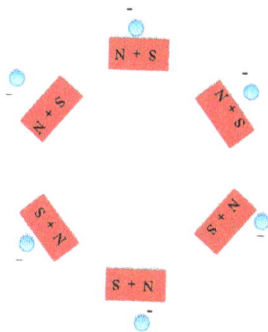

Archion model of atom by J. Stark (1910)

In 1909 year, under Ernest Rutherford direction, the German physicist **Hans Wilhelm Geiger** (1882-1945) and the New Zealand physicist **Ernest Marsden** (1889-1970) bombarded a sheet of gold foil with alpha rays and discovered that a small percentage of these particles were deflected through much larger angles. Rutherford interpreted the gold foil experiment as suggesting that the positive charge of a heavy gold atom and most of its mass was concentrated in a nucleus at the centre of the atom.

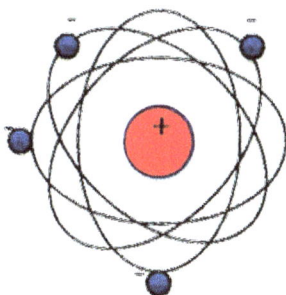

Planetary model of atom by E. Rutherford (1911)

In 1911 year Rutherford theorized that atoms have their positive charge concentrated in a very small nucleus surrounded by orbiting electrons and thereby pioneered the *planetary model* of

91

atom whose structure is shown in figure, alike to the sun surrounded by the orbiting planets. He considered that nearly all the atom was nothing but an empty space.

The Rutherford planetary model of the atom had two significant shortcomings.

The first shortcoming of the Rutherford model of atom was that, unlike planets orbiting a sun, electrons are electrically charged particles and the motion of the electrons in the Rutherford model was unstable because according to classical mechanics and electromagnetic theory, any charged particle moving on a curved path emits electromagnetic radiation, so the electrons would lose energy and fall into the nucleus.

The second shortcoming of the Rutherford model of atom was that the planetary model like all other models could not explain the highly peaked emission/absorption spectra of atoms.

The Rutherford model of atom was rendered irrelevant by the Danish physicist and philosopher **Niels Henrik David Bohr** (1885-1962) in year 1913 when, to remedy the stability problem, Bohr proposed his atom *model with quantized shell orbits,* a *monumental discovery,* explaining how electrons can have stable orbits around nucleus. Bohr modified the Rutherford model by requiring that the electrons move in circular orbits of fixed size and energy at a certain set of distances from the nucleus (the orbits are quantized) and electrons do not continuously lose energy as they travel. The only orbits that are allowed are those for which the angular momentum of the electron is an integral multiple of h/2 where h is the Planck constant.

Angular momentum is a vector quantity, requiring the specification of both, magnitude and direction for its complete description. With the direction of the rotation axis, the magnitude of the angular momentum L of an orbiting object of mass m and linear velocity v is: L = *mvr* = pr

p = mv = linear momentum of object

r = perpendicular distance from the centre of rotation to a line drawn in the direction of its instantaneous motion and passing through the object's centre of gravity

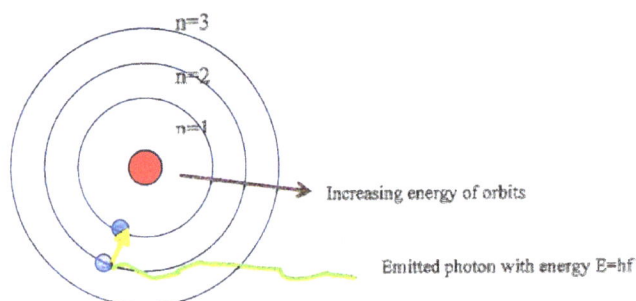

Quantized shell model of atom by N. Bohr (1913) - WINNER!

Bohr's idea was that each discrete orbit could only hold a certain number of electrons. After that orbit is full, the next level would have to be used. This gives the atom a *shell structure*, in which *each shell corresponds to a Bohr orbit*. The energy of an electron depends on the size of the Bohr orbit and is lower for smaller orbits. The electrons can only gain and lose energy by jumping from one allowed Bohr orbit (shell) to another, absorbing or emitting electromagnetic radiation with a frequency f determined by energy difference E of the levels according to Planck relation E=hf. Bohr's model of the atom is important because it introduced the concept of quantum in explaining atomic properties. It was considered the winning model and for his services in the investigation of the structure of atoms, Niels Bohr was awarded the **1922 Physics Nobel Prize**.

So after the *cubic model* (1902), *dynamid model* (1903), *plum-pudding model* (1904), *Saturnian model* (1904), *electron fluid model* (1906), *vibrating electron model* (1906), *expanding electron model* (1906), a*rchion model* (1910) and *planetary model* (1911) came the *quantized shell model* (1913), which was the WINNER.

The Bohr model of atom was an improvement on older atomic models, but it too has been rendered obsolete by ongoing scientific research. The Bohr model ran into problems with heavier than hydrogen atoms, failing to explain much of the spectra of larger atoms or giving an incorrect value for the ground state of orbital angular momentum.

The French physicist **Prince Louis-Victor-Pierre-Raymond, Seventh Duc de Broglie** (1892-1987) proposed in 1924 year that all moving particles, particularly subatomic particles such as electrons, exhibit a degree of wave-like behaviour. This concept is known as *wave-particle duality* or *de Broglie hypothesis*. Traditional, classical physics had assumed that particles were only particles and waves were only waves. However, de Broglie suggested that in nature, the particles could sometimes behave as waves with wavelengths λ and waves could sometimes behave as particles with mass m and velocity v.

He suggested the wave-particle duality defined by the simple equation λ = h/mv.

This wave-particle duality turned out to be a *key to a new atomic theory*.

In fact, de Broglie's theory demands that all moving objects - not just electrons but baseballs, birds, cars - have an associated wavelength, but we cannot observe those effects in the visible world because the wavelength gets smaller as the mass and velocity get larger; a bowling ball rolling down the alley has a wavelength much shorter than any existing device can measure. Louis de Broglie was awarded the **1929 Physics Nobel Prize** for *de Broglie hypothesis* after the wave-like behaviour of matter (electrons) was first experimentally demonstrated in 1927 year by the American physicists **Clinton Josef Davisson** (1881-1958) and

Lester Halbert Germer (1896-1971) in the United States and the English physicist
George Paget Thomson (1892-1975) in Scotland.

The Austrian physicist **Erwin Rudolf Josef Alexander Schrödinger** (1887-1961), fascinated by Louis de Broglie's idea, explored whether or not the movement of an electron in an atom could be better explained as a wave rather than as a particle.

De Broglie's wave hypothesis formed the basis for the *Schrödinger's equation*, formulated in 1925 year and published in 1926 year. It describes an electron as a wave instead of a point particle: $H\Psi = E\Psi$ Ψ = Orbital   E = Energy   H = Hamiltonian

(The Hamiltonian represents the total energy of system, sum of kinetic and potential energy)   A wave function for an electron in an atom is called an *atomic orbital*.

In the standard interpretation of Quantum Mechanics, the wave function is the most complete description that can be given to a physical system. Solutions to Schrödinger's equation can describe molecular, atomic and subatomic systems, also macroscopic systems, possibly even the whole universe. Schrödinger equation, in its most general form, is consistent with both classical mechanics and special relativity, but the original formulation by Schrödinger himself was non-relativistic. Erwin Schrodinger was awarded **1933 Physics Nobel Prize** for the formulation of Schrödinger equation.

In 1925 year the Austrian-Swiss physicist, pioneer of Quantum Physics **Wolfgang Ernst Pauli** (1900-1958) in order to resolve inconsistencies between observed molecular spectra and developing theory of Quantum Mechanics emphasized that *no more than two electrons can occupy the same orbital* and that *two electrons in the same orbital must have opposite spin angular momentum (spin), spinning in only one of two directions, sometimes called* up *and* down, statement known as the *Pauli principle* or *exclusion principle*.

Spin is an intrinsic form of angular momentum carried by electrons and also other particles and atomic nuclei (around own axis). All elementary particles of a given kind have the same magnitude of spin angular momentum, which is indicated by assigning the particle a *spin quantum number s*, unitless number. Wolfgang Ernst Pauli was awarded the **1945 Physics Nobel Prize** for his decisive contribution through his discovery of a new law of Nature, the *exclusion principle* or *Pauli principle*.

Although Erwin Schrodinger concept was mathematically convenient, it was difficult to visualize and faced opposition. One of its critics, the German physicist and mathematician **Max Born** (1882 1970), proposed in 1926 year that Schrödinger's wave function Ψ describes not the electron but rather all its possible states, giving the probability to find an electron at   any location around nucleus.

This reconciled the two opposing theories of particle versus wave electrons and the idea of wave-particle duality was reintroduced, the electron may exhibit the properties of both, wave and particle. Max Born won the **1954 Nobel Prize in Physics** especially for his fundamental research in the statistical interpretation of the wave function.

The unfamiliar world of the unimaginably small, where the rules of physics seem to be different from the rules in the world we can see and touch was discovered through the scientists' attempts to understand the structure of atom. Mathematics is used to explore this world allowing the scientist to explore beyond the boundaries of the world we can experience directly.
The modern model of the atom describes the positions of electrons in an atom in terms of probabilities. An electron can potentially be found at any distance from the nucleus, but depending on its energy level, exists more frequently in certain regions around the nucleus than others. The best way to imagine the electrons is a 'cloud".

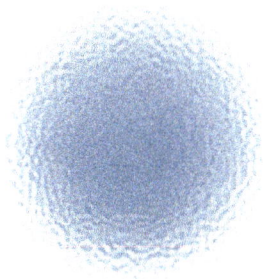

View of the cloud of electrons around the nucleus

The electrons cloud represents the continuous variation in intensity of the negative electric charge, stronger near the nucleus and weaker with increasing distance away from nucleus.
At present the theoretical model of the atom involves a dense nucleus surrounded by a probabilistic *cloud* of electrons.
Each electron seems to have a dual nature in which both particle and wave characteristics are apparent. It is difficult to describe these two aspects of an electron at the same time and so sometimes the focus is on its particle nature and sometimes on its wave character.
Each electron in atom is described by *four different quantum numbers: n, l, m, s* .

The first three quantum numbers *n*, *l*, *m* obtained by solving the Schrödinger equation specify the particular orbital of interest and the fourth *s* specifies how many electrons can occupy that orbital since Pauli principle/exclusion principle. Electron symbol is "e–".

Electron is a subatomic, elementary, negatively electric charged particle with electric charge –1e, spin quantum number 1/2, orbit radius about 1 Å$=10^{-10}$ m and mass about 9.109×10^{-31} kg or 0.511 MeV/c^2.

Mass equivalent to energy of 1eV is 1.783×10^{-36} kg.

Electric charge of 1e is $160.217657 \times 10^{-21}$ C.

Electron, in the wave view, can be described as a three-dimensional standing wave of certain orbitals with varying negative electric charge intensity at different positions outside the nucleus.

NEW SUBATOMIC PARTICLES

Physics - the natural science that involves the study of Matter and its motion through space and time, along with related concepts such as Energy and Force - and *Mechanics* - the area of science, branch of Physics, concerned with the behaviour of physical bodies when are subject to forces or displacements and the subsequent effects of the bodies on their environment - appear to have different laws at atomic and subatomic length scales, scales typical for the quantum of action.

At that level they are named *Quantum Physics* and *Quantum Mechanics*.

Physics and Mechanics are renamed *Classical Physics* and *Classical Mechanics*.

Classical Physics remains an excellent science for bodies and systems at macroscopic scale.

It is the oldest science with centuries of development by giants of reasoning ranging from antiquity to our modern times.

But the behaviour of electrons in atoms and molecules and the nature of chemical bond, are among the problems that classical physics is unable to describe. Only following the development of the quantum theory, the chemists could really use physical ideas to provide a satisfactory understanding of those problems.

Ideas that some material particles consist of smaller particles and that there exists a limited number of sorts of primary, smallest particles, "elementary" or "fundamental" particles in the nature existed in natural philosophy since the time immemorial. Such ideas gained physical credibility since 19th century, but the concept of "elementary" or "fundamental" particle underwent some changes in its meaning. Modern physics does not deem elementary particles un-destroyable any more. Even elementary/fundamental particles can decay or collide destructively. They can cease to exist and create other particles in result.

Small particles in the hierarchy of the Universe were discovered and researched, like molecules that are constructed of atoms that in turn consist of subatomic particles, namely atomic nuclei and electrons. But many more types of subatomic particles were found. Most of them eventually appeared to be composed of even "smaller" particles.

Particle Physics now studies these smallest particles, their properties and behaviour under high energies. Theoretical and experimental research in Particle Physics has given to scientists a clearer understanding of the nature of matter and energy and of the origin of the universe.

Modern Particle Physics began in 1930 years as a result of experimental studies of nuclear and cosmic ray physics being carried out with improved techniques as electronic counters or particle accelerators.

Nuclear Physics studies atomic nuclei and their immediate constituents.

As physicists penetrated deeper into the small nucleus, they encountered the universe by learning about thermonuclear reactions in sun, about producing new and strange forms of matter facing the creation of the elements in universe.

Subatomic Physics deals with all entities smaller than the atom, combining the nuclear and particle physics. Subatomic physics has enriched science with new concepts and deeper insights into the laws of nature. More than 200 subatomic particles have been detected - most of them highly unstable, existing for less than a millionth of a second - as a result of collisions produced in cosmic-ray reactions or particle-accelerator experiments.

Completing the revolution that began in the early 20th century with theories of the equivalence of Matter and Energy, the study of subatomic particles has been transformed by the discovery that the actions of Forces are due to the exchange of "force particles".

The exploration of Subatomic Physics started by serendipity in 1896 year when Henri Becquerel discovered the uranium rays, followed by Marie Curie research and discoveries in uranium radioactivity and has continued since with the research work of many scientists. It has been a constant source of surprises, unexpected phenomena and fresh insights into the laws of nature.

In 1907 year, Ernest Rutherford proved that alpha radiation was helium ions.

In 1917 year he appeared again in focus, *first splitting the atom in a nuclear reaction* between nitrogen and alpha particles, when he had discovered that the hydrogen nucleus, known to be the lightest nucleus, could be extracted from the nuclei of nitrogen by collision.

In 1920 year Ernest Rutherford gave the name "proton" to the hydrogen nucleus.

The word "proton" is the Greek word for "first".

Actually the hydrogen nucleus is the subatomic particle *proton* with positive electric charge +1e, mass approximately one atomic mass unit and symbol p or p+.

Classical Physics assumes that precise location and velocity of objects is always possible. However the German theoretical physicist **Werner Karl Heisenberg** (1901-1976) discovered that this was not necessarily the case at the atomic level.

Heisenberg's uncertainty principle or the *uncertainty principle,* introduced first in 1927 year by Werner Heisenberg, is any of a variety of mathematical inequalities *asserting a fundamental limit to the precision with which certain pairs of physical properties of a particle known as complementary variables can be known simultaneously.*

Examples of complementary properties are: position and momentum, spin on different axis, wave and particle, value of a field and its change.

For the complementary variables position and momentum, it states that the more precisely the position of a particle is determined, the less precisely its momentum can be known and vice versa.

The uncertainty principle is inherent in the properties of all wave-like systems and it arises in Quantum Mechanics simply due to the matter-wave nature of all quantum objects. It has important implications to what we can measure at the atomic level.

Werner Heisenberg won **1932 Physics Nobel Prize** for creation of Quantum Mechanics.

In 1930 year, the English physicist **Sir John Douglas Cockcroft** (1897-1967) and Irish physicist **Ernest Thomas Sinton Walton** (1903-1995) built, at Cavendish laboratory in Cambridge England, the 800 kilovolts voltage multiplier, a stack of capacitors connected by rectifying diodes as switches, supplied by a 200 kilovolts transformer. They used the voltage to accelerate protons down a void air, eight feet long tube, to propel the protons to high speeds containing them in well-defined beams along a straight line, since they wanted a test for a phenomenon known as "Gamow's tunnelling". This was the *first particle accelerator machine.*

The accelerators speed up and increase the energy of a beam of particles by generating electric fields that accelerate particles and magnetic fields that steer and focus them.

There are two basic classes of accelerators: *electrostatic* and *oscillating field*.

There are two types of accelerators: *linear accelerators* that accelerate particles over a long, straight line and *circular accelerators* where strong magnets bend the particle's path into a circle.

In modern accelerators the particles very quickly reach almost light speed of 299,792.458 km/s.

The highly energetic particles in an accelerator can be used for particles fundamental research.

The particles are subatomic particles like electrons or protons, or - in special experiments - even whole atoms (heavy ions). As a result of their high energy, these particles are not only scattered but also transformed into other particles. Higher energies demand for particles accelerators continues to this day. Particle accelerators are commonly referred to as "atom smashers". Modern particle accelerators are gargantuan compared with their forerunners.

Large particle accelerators are best known for their use in Particle Physics as "colliders", for example:

- Tevatron 1987-2011 at Fermilab (Fermi National Accelerator Laboratory)

- Relativistic Heavy Ion Collider RHIC 2000 at BNL (Brookhaven National Laboratory)

- Large Hadron Collider LHC 2009, circular accelerator, circumference 27km at CERN

CERN is the acronym of 'Conseil Européen pour la Recherche Nucléaire' respectively 'European Council for Nuclear Research'.

There are currently more than 30,000 accelerators in operation around the world, because the experiments in Nuclear Physics and Particle Physics have to be carried out, with very few exceptions, using particle accelerators.

The development and construction of accelerators with even greater energies and beam intensities have made possible to discover more and more elementary particles.

The unit of energy used in this field is the "electron-volt" symbol "eV" (not SI unit), defined as the amount of energy gained or lost by the electric charge of a single electron 1e, moved across an electric potential difference of one volt 1V.

"electron-volt" or "eV" is a unit of energy equal to $\sim 1.6 \times 10^{-19}$ joules (J).

$1V=1$ J/C $e=1.602176565(35) \times 10^{-19}$ C 1 eV$=1V \times 1e=1.602176565(35) \times 10^{-19}$ J

The mass of proton is equivalent almost exactly to the energy of 1 GeV=1 billion eV.

The construction of powerful particle accelerators enabled the study of high-energy collisions of electrons, protons and other particles. The most modern and largest accelerator of the world, LHC reached 13 TeV total in 2015 year. As particles collide at high energy, the collision energy becomes available for the creation of subatomic particles such as mesons and hyperons. Through conversion of energy into mass, any particle can be produced through collision of other particles at high energy.

Paul Adrien Maurice Dirac (1902-1984) English theoretical physicist, one of the pioneers of Quantum Mechanics, found in 1931 year an unexpected minus sign in the equations he was developing to describe the nature of matter. He suggested that the minus sign indicated the existence of a particle identical to the electron except that it has a positive electric charge instead of a negative one. In 1932 year, this "anti-electron" was detected in cosmic rays by the American physicist **Carl David Anderson** (1905-1991) and called "positron" with symbol "e+".

Carl David Anderson was awarded **1936 Physics Nobel Prize** for his discovery of the *positron* shared with Austrian-American physicist Victor Franz Hess for the discovery of cosmic radiation. *Positron* or *antielectron* is the antiparticle or the antimatter counterpart of the electron, with the same mass as an electron, spin quantum number of 1/2 and an electric charge +1e. When a low-energy positron collides with a low-energy electron, annihilation occurs, resulting in the production of two or more gamma ray photons. Positrons may be generated by positron emission radioactive decay or by pair production from a sufficiently energetic photon, which is interacting with an atom in a material. Carl Anderson first detected the positrons in cosmic rays. He then produced more conclusive proof by shooting gamma rays produced by a natural radioactive nuclide into other materials, resulting positron-electron pairs.

Apparently, every particle has a twin antiparticle or antimatter counterpart and they are formed together from very concentrated energy.

Antimatter is material composed of antiparticles, which have the same mass as particles of ordinary matter but have opposite electric charge. Or they may also be opposite in other properties, including spin quantum number and colour charge. An unsolved mystery of cosmology is why matter rather than antimatter dominates the universe because the universe seems to be made up entirely of particles and that the antiparticles are so rare.

There are theories that particles and antiparticles ought to exist in the universe in equal quantities. Paul Dirac predicted the existence of "antimatter" and other particle properties relaying on the symmetry between positive and negative energy.

Among other discoveries, he formulated the "Dirac equation", which describes the behaviour of the particles called "fermions", particles with half-integer spin quantum number, and did work that forms the basis of modern attempts to reconcile Quantum Mechanics with *general* theory of *relativity*, the current description of gravitation in modern physics.

Paul Dirac was awarded **1933 Physics Nobel Prize** for the discovery of new productive forms of atomic theory, shared with Erwin Schrödinger.

Before 1920 year, the "father of nuclear physics" Ernest Rutherford and other scientists observed a difference between the atomic number of an atom, which is the number of its positive charges (number of protons) and the atomic mass computed in atomic mass units.

Rutherford suggested that the disparity could be explained by a neutrally charged particle within the atomic nucleus, the "neutron".

In 1931 year, the German nuclear physicists **Walther Bothe** (1891-1957) and **Herbert Becker** in Giessen, Germany found that if the very energetic alpha particles emitted from polonium element fell on certain light chemical elements, beryllium, boron or lithium, an unusually penetrating radiation was produced. This radiation was not influenced by an electric field.

Light elements are the chemical elements produced during the first few minutes after the 'Big Bang', including, in addition to hydrogen and helium, lithium, beryllium and boron. The 'Big Bang' is the accepted model of our Universe formation in which the space-time and the matter within it were created from a cosmic singularity. The model suggests that in the 13.7 billion years since the Universe began, it has expanded from an extremely small, hot and dense primordial point to the huge, cold and diffuse Universe of today.

Further expansion 'Open Universe' or contraction 'Big Crunch' / 'Closed Universe' depend on the present mass of Universe. Inflationary Universe is that critically poised between the two, with a critical mass allowing a continuous but slow expansion.

In the year 1932 in Paris, the French scientists **Irène Curie Joliot** (1897-1956) and **Frédéric Joliot** (1900-1958) showed that when the unknown radiation fell on paraffin or any other hydrogen-containing compound, it ejected protons of very high energy. In Rome, the young Italian theoretical physicist **Ettore Majorana** (1906 -1938) was the first to correctly interpret the experiment as requiring a new particle of neutral charge and a mass about the same as the proton.

The English physicist **Sir James Chadwick** (1891-1974) studied under the supervision of Ernest Rutherford until 1913 year. After the war, Chadwick followed Rutherford to the Cavendish Laboratory at the University of Cambridge where he continued to research for Rutherford's "neutron" throughout the 1920s without success. He repeated the creation of the radiation using beryllium, used better approaches to detection and aimed the radiation at paraffin following the Paris experiment. Paraffin is high in hydrogen content, a target dense with protons and Chadwick measured the range of the scattered protons. He also analysed the impact of radiation on the atoms of various gases. He found in 1932 year that the new radiation consisted of uncharged particles with about same mass as protons, the "neutrons".

James Chadwick won **1935 Nobel Prize in Physics** for this discovery of the *neutron*.

Neutron, symbol "n", is a subatomic particle with no net electric charge and a mass slightly larger than that of a proton. Comparatively the masses of neutron, proton and electron are:

In kilograms: Neutron $1.6749275 \times 10^{-27}$ kg

Proton $1.6726217 \times 10^{-27}$ kg

Electron $9.1093829 \times 10^{-31}$ kg

Relative to neutron: Neutron   1

   Proton   0.99862349

   Electron   0.00054386

In atomic mass units: Neutron   1.00866 u

   Proton   1.00727 u

   Electron   0.00054 u

In terms of MeV/c^2: Neutron   939.5654 MeV/c^2

   Proton   938.2723 MeV/c^2

   Electron   0.51099 MeV/c^2

The "unified atomic mass unit" with symbol "u", called also "Dalton" with symbol "Da" is the standard unit that is used for indicating mass at atomic or molecular scale: u = 1.66 x 10^{-27} kg

Neutrons and protons, each with mass approximately one atomic mass unit u, constitute the nucleus of an atom and they are collectively referred to as "nucleons" and their properties and interactions are described by the Nuclear Physics.

The nucleus consists of a number of protons, called "atomic number" with symbol Z and a number of neutrons, called "neutron number" with symbol N.

The "atomic mass number" A with symbol A equals Z+N.

The *atomic number Z* defines the chemical properties of the atom or element and the *neutron number N* determines the "isotope" or "nuclide" of the element.

The terms "isotope" or "nuclide" are often used synonymously, but they refer to chemical or nuclear properties respectively.

Within the nucleus, protons and neutrons are bound together through the nuclear force and neutrons are required for the stability of nuclei. Neutrons are produced copiously in nuclear fission and fusion. They are a primary contributor to the nucleosynthesis of chemical elements within stars, through fission, fusion and neutron capture processes.

The neutron is essential to the production of nuclear power.

So in 1932 year Rutherford's colleagues had known that the nucleus is built of smaller particles, the positive electric charged *protons* and the neutral electric charged *neutrons*. Only after the discovery of neutron was achieved a real understanding of the particles that compose the nucleus of an atom and given a *real definition of the radioactivity* as a *phenomenon resulting from an instability of the atomic nucleus in certain atoms whereby the nucleus experiences a spontaneous but measurable delayed nuclear transition or transformation with the resulting emission of radiation.*

All atoms of a given element have a given number of protons in their nuclei called the *atomic number* defining the element. To balance this charge they have an equal number of electrons swarming around the nucleus in shells of electrons that give the element its chemical properties. It turned out that atoms of a given element, with a given number of protons, can have different numbers of neutrons and Frederick Soddy was the first to conclude in 1913 year that the radioactive elements might exist in forms that differ in atomic mass while being indistinguishable and inseparable chemically with the same atomic number, those forms being the *isotopes* of those elements. Explorations in the composition of canal rays (positive ions) by J.J. Thomson in 1913 year, led to the first evidence that non-radioactive elements can also have multiple isotopes. Example is the lightest element, hydrogen.

Hydrogen has the atomic number $Z=1$, its nucleus is normally made of one proton and no neutrons and thus its atomic mass number is also 1, $A=1$. But hydrogen has isotopes with different atomic mass numbers; "heavy" hydrogen, called "deuterium", has one proton $Z=1$ and one neutron $N=1$ in its nucleus, thus its atomic mass number is $A=2$; hydrogen also has a radioactive isotope called "tritium" that has one proton $Z=1$ and two neutrons $N=2$, thus its atomic mass number $A=3$.

The three forms of hydrogen each have one electron and thus the same chemical properties. When a radioactive nucleus gives off alpha particles, it is in the process of changing into a different nucleus; that can be a different element or a different isotope of the same element.

For example, radioactive thorium is formed when uranium 238 or U-238, an isotope of uranium with 92 protons $Z=92$, 146 neutrons $N=146$ and atomic mass number $A=92+146=238$, emits an alpha particle. Since the alpha particle consists of two protons $Z=2$ and two neutrons $N=2$, when these are subtracted what is left is a nucleus with 90 protons $Z=90$ and 144 neutrons $N=144$, corresponding to a new element of $A=90+144=234$, an isotope of thorium, thorium 234, Th-234. Thorium is the element of atomic number 90 and this isotope of thorium has the atomic mass number 234. The results of decay may themselves be unstable, as is the case with thorium 234. The chain of decays continues until a stable nucleus forms, in this case the element lead.

Ernest Rutherford and Frederick Soddy discovered that half the nuclei in a given quantity of a radioactive isotope will decay in a specific time period, meaning that every radioactive isotope has a *specific half-life*. The half-life of uranium 238 is 4.5 billion years. The isotopes produced by the decay of uranium, themselves decay in a long chain of radiations. The elements radium and polonium are links in this chain.

Radium caught Marie Curie's attention because its half-life is 1600 years, long enough to be a fair amount of radium mixed with uranium in her pitchblende, and also short enough its radioactivity to be quite intense. A long-lived isotope like uranium 238 emits radiation so slowly that its radioactivity is scarcely noticeable. By contrast, the half-life of the longest-lived polonium isotope, polonium 210 or Po-210 is only 138 days, explaining why Marie Curie was unable to isolate polonium. Even she performed her meticulous fractional crystallizations, the polonium in her raw material was disappearing as a result of its rapid radioactive decay.

The lifetime or mean lifetime or exponential time constant τ for a decaying particle relates to the decay rate λ by the relation $\tau = 1/\lambda$ and the time for a decaying quantity to fall at half of its initial value is its half-life $t_{1/2} = \ln2/\lambda = \tau\ln2 = 0.693\ \tau$.

In the year 1934, from theoretical considerations, the Japanese theoretical physicist **Hideki Ogawa Yukawa** (1907-1981) predicted the existence of the particle "meson" as the carrier of the nuclear force that holds atomic nuclei together. Otherwise all nuclei with two or more protons, which are electrically positive, would fly apart because the protons would reject each other. Yukawa called his carrier particle "meson", from μέσος (mesos), the Greek word for "intermediate," because its predicted mass was between that of the electron and that of the proton. The lightest and more stable meson with a lifetime of $2.6\ \times10^{-8}$ seconds, later called "pi meson" or "pion", was first discovered in 1947 year by the physicists British **Cecil Powell** (1903-1969), Brasilian **César Lattes** (1924-2005) and Italian **Giuseppe Occhialini** (1907-1993), who were investigating cosmic ray products at the University of Bristol in England, based on photographic films placed in the Andes Mountains. Over the next few years, more experiments showed that the "pi meson" was indeed the "primary force carrier" for nuclear force in atomic nuclei; other mesons such "rho meson" are also involved but to lesser extents.

The mesons are not produced by radioactive decay, but appear in nature only as short-lived products of very high-energy interactions in matter, between particles made of "quarks".
In cosmic ray interactions such particles are ordinary protons and neutrons. Mesons are also frequently produced artificially in high-energy particle accelerators that collide protons, anti-protons or other particles.

Meson has a physical size with a diameter roughly 1fm, which is about 2/3 the size of a proton or neutron, mass $139\text{MeV/c}^2 - 9.460\ \text{GeV/c}^2$, spin quantum numbers 0 or 1 and an electric charge of -1e, 0e, +1e.

Hideki Ogawa Yukawa was awarded the **1949 Physics Nobel Prize** for *meson* prediction after the experimental discovery of the *pi meson.*

In 1936 year Carl David Anderson and his first graduate student,
Seth Henry Neddermeyer (1907-1988) discovered in cosmic rays again, the "muon", the first of a long list of believed to be subatomic, elementary/fundamental particles - particles which are not believed to have any sub-structure, not thought to be composed of any simpler particles.

Muon is an elementary/fundamental particle, classified as charged lepton, with negative electric charge of -1e and spin quantum number s of 1/2 as the electron, but 207 times more massive than the electron with mass of 105.7 MeV/c^2 and is an unstable subatomic particle with a lifetime of 2.2 μs. Muon decay always produces at least three particles, including a same charge electron and two neutrinos of different types. Muons are denoted by the symbol μ−.

The muon has a corresponding antiparticle of opposite electric charge but equal mass and spin, the *antimuon* or the *positive muon*. Antimuons are denoted by the symbol μ+.

After the neutron was discovered in 1932 year, it was quickly realized that neutrons might act to form a nuclear chain reaction. The nuclear fission was discovered in 1938 year and became clear that, if a fission event produced neutrons, each of these neutrons might cause further fission events in a cascade known as a chain reaction.

The early 20th-century explorations of Nuclear Physics and Quantum Physics culminated in proofs of *nuclear fission* in 1939 year by Austrian physicist **Lise Meitner** (1878-1968), based on experiments by German chemist **Otto Hahn** (1879-1968) and *nuclear fusion* by the German-American nuclear physicist **Hans Albrecht Bethe** (1906-2005) in that same year.

Both discoveries also led to the development of nuclear weapons. These events and findings led to the first self-sustaining nuclear reactor Chicago Pile-1 in 1942 year and the first nuclear weapon Trinity in 1945 year.

The individual neutrons free of the nucleus, the *free neutrons,* are a form of ionizing radiation and depending upon dose a biological hazard. Caused by cosmic rays and by the natural radioactivity of spontaneously fissionable elements in the Earth's crust, a small natural "neutron background" flux of free neutrons exists on Earth. Dedicated neutron sources like neutron generators, research nuclear reactors and spallation neutron sources - the accelerator-based facilities providing the most intense pulsed neutron beams in the world - produce free neutrons for use in irradiation and in neutron scattering experiments.

In 1956 year the American physicist **Leon Cooper** first described how, in condensed matter, a pair of electrons is bound together at low temperatures in a certain manner as a "quasiparticle" called *Cooper pair* or *BCS pair.*

Quasiparticle, in physics is considered a disturbance that behaves as a particle and may be conveniently regarded as one.

Cooper showed that an arbitrarily small attraction between electrons in a metal could imply that the pair is bound. In conventional superconductors, this attraction is due to "electron-phonon" interaction. *Phonon* is a unit of vibrational energy arising from oscillating atoms in a crystal. The *Cooper pair state* is responsible for superconductivity as described in the BCS theory, the first microscopic theory of superconductivity since its discovery in 1911 year.

For developing the BCS theory **John Bardeen**, **Leon Cooper** and **John Schrieffer** were awarded the **1972 Physics Nobel Prize**.

In 1956 year the American physicists **Clyde Lorain Cowan** (1919-1974) and **Frederick Reines** (1918-1998) discovered experimentally another subatomic elementary/fundamental particle, identified as "neutrino", particle already proposed theoretically by the physicist Wolfgang Ernst Pauli in December 1930 year because he tried to explain why the amount of energy that an atom give off with beta particles is less than that scientific theories demand. The word "neutrino" means "little neutral one" in Italian language.

Neutrino is a weakly interacting subatomic, elementary/fundamental particle with half-integer spin s =1/2. Its symbol is denoted by the Greek letter ν (*nu*).

All evidence suggests that neutrinos have mass but that their masses are tiny, even by the standards of subatomic particles. Neutrinos do not carry any electric charge, which means that they are not affected by the electromagnetic force that acts on charged particles, such as electrons and protons. Neutrinos are affected only by the weak subatomic force, which is of much shorter range than electromagnetism or gravity and gravity is too weak at subatomic scale.

Therefore, a neutrino typically passes through normal matter unimpeded.

Neutrinos have three generations: *electron neutrinos, muon neutrinos, tau neutrinos*

Electron neutrinos were predicted in 1930 year by Wolfgang Pauli and discovered in 1956 year by Clyde Cowan and Frederick Reines.

Muon neutrinos were predicted in 1940 year and discovered in 1962 year by the American physicists **Leon Max Lederman, Melvin Schwartz** (1932-2006) and **Hans Jakob Steinberger**.

Tau neutrinos discovery was announced in 2000 year by DONUT (Direct Observation of NU Tau) collaboration.

Each generation is associated with an antiparticle, called *antineutrino*, which also has neutral electric charge and half-integer spin. The problem of whether or not the neutrino and its corresponding antineutrino are identical particles has not yet been resolved.

The solar neutrino flux on Earth is about *65 billion neutrinos* passing through one square centimetre of area on earth *every second*, through the earth and out the other side.

That means over the course of our lifetime, about 10^{23} neutrinos will stream through our body. Leon Lederman, Melvin Schwartz and Hans Steinberger received **1988 Nobel Prize in Physics** for their research on neutrinos and Frederick Reines was awarded the

1995 Nobel Prize in Physics for the experimental discovery of the electron neutrino.

Throughout the 1950s and 1960s, many particles were found in "scattering" experiments. It was referred to as the "particle zoo", informal term in Particle Physics used to describe the known group of new particles that almost look like hundreds of species in the zoo. "Scattering" refer to particle-particle collisions between molecules, atoms, electrons, photons and other particles. The scattering experiments were an extension of "Rutherford scattering", to much higher energies of the scattering particles.

Experiments at particle accelerators in the 1950 and 1960 years, showed that protons and neutrons are merely representatives of a large family of particles now called "hadrons". By 2008 year more than 200 hadrons, sometimes called the "hadronic zoo", have been detected. The scattering of electrons allowed to understand that protons and neutrons are composite particles made up of smaller subatomic, elementary/fundamental particles called "quarks", that they are "composites of quarks" held together by a strong force.

A "composite" particle is a particle containing two or more elementary/fundamental particles. The "deep inelastic scattering experiments" at Stanford Linear Accelerator Centre, SLAC in 1968 year were the first convincing evidence of the reality of "quark" particle, predicted in 1964 year by the American physicists **Murray Gell-Mann** and **George Zweig** independently, which up until that point had been considered by many to be a purely mathematical phenomenon. The name "quark" comes from a poem by Irish writer James Joyce published in 1939 year in the novel "Finnegan's Wake".

Murray Gell-Mann was awarded **1969 Physics Nobel Prize** for his contributions and discoveries concerning the classification of elementary particles and their interactions.

The different varieties (species) of an elementary/fundamental particle are commonly called "flavours", conventionally parameterized with "flavour quantum numbers".

In Particle Physics they are assigned to all subatomic particles, including composite ones. For hadrons, these flavour quantum numbers depend on the numbers of constituent quarks of each particular flavour.

Quarks come in six flavours. In order of increasing mass, they are referred to as 'up', 'down', 'strange', 'charm', 'bottom' and 'top'. For each of these quarks, there is an *antiquark*. Quarks come in three generations: 'up/down', 'charm/strange', 'top/bottom'.

The *up, down, strange quarks* were postulated in 1964 year by Murray Gell-Mann and George Zweig and first observed by experiments in 1968 year at SLAC.

The *charm quark* was theorized by the physicists American **Sheldon Lee Glashow**, Greek **John Iliopoulos** and Italian **Luciano Maiani** in 1970 year and discovered in 1974 year in electron-positron annihilation at SLAC by the American physicist **Burton Richter** team and at Brookhaven National Laboratory BNL by the American physicist **Samuel Ting** team.

The *bottom quark* was theorized in 1973 by Japanese physicist **Makoto Kobayashi** and theoretical physicist **Toshihide Maskawa** and was discovered in 1977 year by the team led by Leon Max Lederman in proton collisions at Fermi National Accelerator Lab, Fermilab.

The most massive quark, with the mass of a silver atom, the very instable *top quark* was theorized in 1973 year by the Makoto Kobayashi and Toshihide Maskawa, but it was not observed or discovered till 1995 year in accelerator experiment on proton-antiproton annihilation at Fermilab.

Quark is the particle characterised by mass, spin quantum number, electric charge and also another type of charge called "colour charge".

The colour charge has three manifestations: "red", "blue" and "green". The colour charge manifestations for the antiquarks are termed: "anti-red", "anti-blue" and "anti-green".

Just as combining positive and negative electric charges results in a neutral electric charge, combining colour charges with anti-colour charges gives a neutral colour charge (colourless).

Quarks have spin quantum number 1/2. Quarks' mass is between 2.3 – 173340 MeV/c^2.

Quarks have a fractional electric charge, which are respectively:

up +2/3e; down -1/3e; charm +2/3e; strange -1/3e; top +2/3e; bottom -1/3e

Particles carrying a colour charge, such as quarks, cannot exist in free form, property called "colour confinement" meaning that colour charged particles cannot be isolated singularly and therefore cannot be directly observed.

A top quark has a lifetime of 5×10^{-25} seconds, which is shorter than the time scale at which the strong force of QCD acts, so a top quark decays before it can hadronize, allowing physicists to observe a "bare quark." Thus, top quarks have not been observed as components of any hadron, while all other quarks have been observed only as components of hadrons.

The **2008 Physics Nobel Prize** was awarded to Makoto Kobayashi & Toshihide Maskawa "for the prediction of the top and bottom quark, which together form the third generation of quarks" shared with Yoichiro Nambu "for discovery of the mechanism of spontaneous broken symmetry in subatomic physics"

In 1955 and 1956 years, *antimatter* was detected and created as *antiparticles*: *antiprotons* and *antineutrons*. The general name for anti-particles is anti-matter. As the terminology suggests, when a particle meets its antimatter counterpart, they annihilate each other, leaving pure energy in their place (although by elastic scattering may interact with each other without annihilating). As example, when a *positron* particle collides with an *electron* particle at low energy, they both disappear, sending out two *gamma rays* (γ-rays), photon particles, in opposite directions.

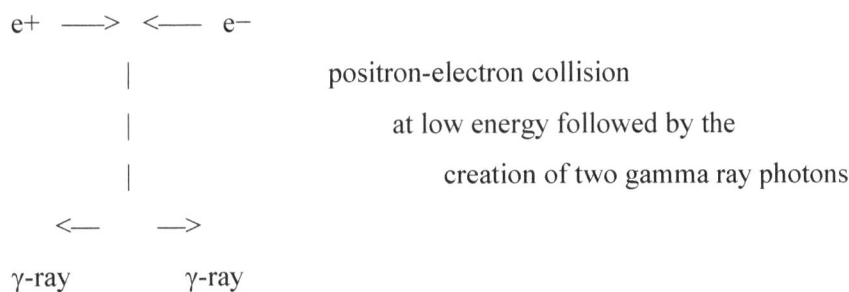

```
e+ ——>  <—— e-
       |              positron-electron collision
       |                  at low energy followed by the
       |                      creation of two gamma ray photons
   <—      —>
 γ-ray      γ-ray
```

Positron emission tomography PET is a technique using antimatter created in the laboratory to scan the brain in search of biochemical abnormalities that signal the presence of Alzheimer's disease, schizophrenia, epilepsy, brain tumours and other brain disorders.

In PET procedure, radioactive substances that emit positrons are introduced into a patient's bloodstream. As the radioactive atoms decay, the positrons they emit collide with electrons, producing gamma rays that escape from the body and are detected by an array of instruments surrounding the patient. Computer analysis of the amount and direction of gamma ray production and comparison of data collected from people, with and without certain brain disorders, provides doctors with valuable information.

Similar experiments are providing information about physiological processes such as glucose metabolism, the effects of opiate drugs and the mechanisms of memory retrieval.

At higher energies the result of the collision between electron and positron is the annihilation of the electron and positron and the creation of other particles. In electron-positron collision at higher energy experiments regularly are observed "two-jet" events. They indicate the annihilation of the two incoming particles and the subsequent creation of a *quark* particle and an *antiquark* particle, both of which result in particle jets.

If the energy of the collision is sufficiently large, either the quark or the antiquark can emit excess energy in the form of the "gluon" particle, which also produces a particle jet in the same plane as the two other jets. Only at high energies, the "gluon jet" appears as a distinct "third jet".

A "jet" is a narrow cone of hadrons and other particles produced by the hadronization of a quark or a gluon in a particle physics or heavy ion experiment.

```
e+ ——> <—— e−
        |            positron-electron collision
        |                at high energy 27.4 GeV, followed by the
        |                    creation of three-jet event: quark, antiquark, gluon
 <——    ——>
quark   |   antiquark
        |
        V  gluon
```

The quantum chromodynamics QCD is the theory of strong interactions between quarks and gluons. The QCD of the hadronization process is not yet fully understood, but it is modelled and parameterized in a number of phenomenological studies, including the Lund string model and in various long-range QCD approximation schemes.

The models and approximation schemes have been extensively compared with measurements in a number of high energy particle physics experiments like TASSO, OPAL, H1.

The physicists Chinese-American **Sau Lan Wu** and American **Georg "Haimo" Zobernig** developed/programmed a method to search for planar three-jet events in TASSO experiment at PETRA collider. At low collision energies, their searches produced no results, but at 27.4 GeV succeeded.

Analysis of the three-jet event showed that, two of the three particle jets were produced by a "quark-antiquark pair" and the third was generated by a "gluon". This collision event, recorded in 1979 year, provided the first evidence of the particle *gluon*, carrier particle of the strong nuclear force, the first direct experimental proof of the existence of elementary/fundamental particle that transmits the strong force, which was theorized by Murray Gell-Mann in 1962 year.

The Norwegian physicist **Bjørn Havard Wiik** (1937-1999) presented this first event at a physics conference in Bergen, Norway.

Gluon is an elementary/fundamental particle, which binds the subatomic particles quarks within the protons and neutrons of stable matter as well as within heavier, short-lived particles created at high energies. Gluon has very tiny mass, spin quantum number 1, no electric charge and carry a strong charge of colour.

The gluons are subject to "colour confinement" meaning that colour charged particles cannot be isolated singularly and therefore cannot be directly observed.

Each of gluons can effectively carry one of eight possible colour/anti-colour combinations resulting eight flavours for gluon. A quark can change colour by emitting or absorbing gluons. All the hadrons (protons, neutrons, mesons) comprised of quarks, antiquarks and gluons have neutral colour charge, are not subject to colour confinement and can exist as independent entities.

In a series of experiments between 1974 -1977 years, the American physicist **Martin Lewis Perl** (1927-2014) and his colleagues at the SLAC-LBL group discovered anomalous events, which indirectly detected the particle "tau". Others subsequently established the mass and spin of "tau".

Tau, also called the *tau lepton, tau particle* or *tauon,* is an elementary/fundamental particle, classified as a charged lepton, similar to the electron with electric charge -1e, spin quantum number 1/2. Tau leptons have a lifetime of 2.9×10^{-13}s and a mass of 1776.82 MeV/c^2 (biggest compared to 105.7 MeV/c^2 for muons and 0.511 MeV/c^2 for electrons).

Tau particle symbol is $\tau-$. Its symbol τ was derived from the Greek word τρίτον (*triton*) meaning "third", since it was the third charged lepton discovered. Tau has a corresponding antiparticle of opposite electric charge but equal mass, the *anti-tau,* also called the *positive tau,* symbol is $\tau+$. Martin Lewis Perl was awarded the **1995 Nobel Prize in Physics** for the experimental detection of *tau,* shared with Frederick Reines for the experimental discovery of neutrino.

The heaviest particle pair yet produced by electron-positron annihilation in particle accelerators is " W^+ boson / W^- boson " pair and the heaviest single particle is the " Z boson ". W^+, W^- and Z bosons are the elementary particles that mediate the weak interaction. Quantum theory enforces the rule that the range of a force varies inversely as the mass of its carrier particle. The gauge bosons of the weak force have to be massive because the weak force has a very small range of action, about 10^{-18} meters.

And because more massive the particle, bigger the accelerator power, the discovery of W and Z bosons had to wait for the construction of a particle accelerator powerful enough to produce them. The first such machine that became available was the Super Proton Synchrotron SPS at CERN where unambiguous signals of "W bosons" were seen in January 1983 during a series of experiments made possible by the Italian particle physicist and inventor **Carlo Rubbia** and the Dutch particle physicist **Simon van der Meer** (1925-2011) with the collaborative effort of many people. "Z boson" was found later in May 1983 year.

W boson has mass 80.385 GeV/c^2, spin quantum number 1, electric charge ±1e and lifetime under 10^{-24} s. W$^-$ is the antiparticle of W$^+$.

Z boson has mass 91.1876 GeV/c^2, spin quantum number 1, electric charge 0e and lifetime under 10^{-24} s.

Carlo Rubbia and Simon van der Meer were awarded the **1984 Nobel Prize in Physics**.

In the early 1960s years, the British theoretical physicist **Peter Ware Higgs** theorised that the "Higgs boson" particle (no spin, no electric charge, no colour charge, decaying into other particles fast) and its associated scalar "Higgs field" were the reason that things have mass.

The Belgian theoretical physicist **François Englert** reached similar conclusions about same time.

On July 4, 2012 the physicists at the CERN Large Hadron Collider LHC announced they have found a new particle that behaves similarly to what is expected from the "Higgs boson".

The composite fermions, protons, were collided at nearly light speed to produce that.

Higgs boson, elementary boson far more massive than proton, mass 125GeV and lifetime 1.56×10^{-22} s with symbol H^0 was confirmed by data from the LHC in 2013 year.

Peter Higgs and Francois Englert shared the **2013 Nobel Prize in Physics** for the prediction of Higgs boson.

LHC is the world's largest and most powerful particle collider built by CERN, Conseil Européen pour la Recherche Nucléaire / European Organization for Nuclear Research, between 1998-2008 years. LHC aim was to allow physicists to test the predictions of different theories of particle physics and high-energy physics and particularly to prove or to disprove the existence of the theorized "Higgs boson" particle, an elementary particle initially theorised in 1964 year and belonging to the large family of new particles predicted by super-symmetric theories.

As of 2015, LHC is the largest and most complex experimental facility ever built.

It took to thousands of scientists, engineers and technicians decades to plan and build and it continues to operate at the very boundaries of scientific knowledge. LHC is expected to address some of the unsolved questions of physics, advancing human understanding of physical laws.

Particles Summary

Particle	Symbol	Antiparticle	Mass MeV/c^2	Spin quantum number	Electric charge e	Colour charge r g b	Lifetime s
Elementary							
Leptons							
electron	e–	e+	0.511	1/2	-1	-	stable
muon	μ–	μ+	105.7	1/2	-1	-	2.2×10^{-6}
tau	τ–	τ+	1776.82	1/2	-1	-	2.9×10^{-13}
electron neutrino	ν_e	own?	$< 0.3 \times 10^{-6}$	1/2	0	-	unknown
muon neutrino	ν_μ	own?	$< 0.3 \times 10^{-6}$	1/2	0	-	unknown
tau neutrino	ν_τ	own?	$< 0.3 \times 10^{-6}$	1/2	0	-	unknown
Quarks	q	q					
up	u	u̲	2.3	1/2	+2/3	yes	stable
down	d	d̲	4.8	1/2	-1/3	yes	900/stable
strange	s	s̲	95.0	1/2	-1/3	yes	1.24×10^{-8}
charm	c	c̲	1290.1	1/2	+2/3	yes	1.1×10^{-12}
bottom	b	b̲	4180.0	1/2	-1/3	yes	1.3×10^{-12}
top	t	t̲	173340.0	1/2	+2/3	yes	5×10^{-25}
Photon	γ	own	$< 1 \times 10^{-12}$	1	$0 / < 1 \times 10^{-35}$	-	10^{18} years/stable
Gluon	g	own	$< 2 \times 10^{-10}$	1	0	yes	stable
W boson	W$^+$	W$^-$	80385	1	+1	-	$<10^{-24}$
Z boson	Z, Z^0	own	91187.6	1	0	-	$<10^{-24}$
Higgs boson	H^0	own	125000.0	0	0	-	1.56×10^{-22}
Graviton theoretical	G	own	0	2	0	-	stable
Composite							
proton	p, p$^+$, N$^+$	p̲	938.2723	1/2	+1	-	10^{32} years/stable
neutron	n, n^0, N^0	n̲	939.5654	1/2	0	-	881.5 (free)
mesons	variety	variety	139.0-9460.0	0, 1	-1, 0, +1	-	10^{-24} - 5×10^{-8}

Note: Data given for orientation

FUNDAMENTAL INTERACTIONS

Interaction is a kind of action that occurs as two or more objects have an effect upon one another. Idea of two-way effect is essential in the concept of interaction, as opposed to one-way causal effect.

The fundamental interactions on which all physical phenomena are based, also known as fundamental forces, are the interactions that do not appear to be reducible to more basic interactions: gravitation, electromagnetic interaction, strong interaction, weak interaction They govern how objects or particles interact and how certain particles decay.

In Particle Physics, each of the four fundamental interactions of Nature - weak interaction, strong interaction, electromagnetism or gravitation - is the manifestation of a "gauge" field, which depends on space-time.

The word "gauge" comes from the term "track gauge" used in rail transport.

On the way from elementary particles to nucleons and nuclei, the fundamental laws of interaction between elementary particles are less and less recognisable in composite systems because many-body interactions cause greater and greater complexity for larger systems.

Each of the fundamental interactions is responsible for Forces in Nature and in consequence is mediated by force carriers particles called "gauge" particles, which belong to the class of particles named "bosons". The interactions are mediated by "gauge bosons".

The four fundamental interactions are presented further.

1. Weak interaction is the mechanism responsible for the *weak force* also called the *weak nuclear force*, cause of radioactive decay of unstable subatomic particles and plays an essential role in nuclear fission.

The force is termed weak because its field strength over a given distance is typically several orders of magnitude less than that of the strong nuclear force or that of the electromagnetic force. The effectiveness of the weak force is confined to a distance range of 10^{-18} metres, less than 0.1% of the diameter of a typical atomic nucleus. The diameter of the nucleus is between 1.6 fm for a proton in light hydrogen to about 15 fm for the heaviest uranium ($1fm=10^{-15}$ m). The diameter of the corresponding atom is bigger by a factor of 145,000 for hydrogen and 23,000 for uranium.

In 1933 year, the Italian physicist **Enrico Fermi** (1901-1954) proposed the "first theory of the weak nuclear interaction", known as *Fermi's interaction* or *Fermi Theory of Beta Decay*. In Nuclear Physics, *Beta Decay* (β decay) is a type of radioactive decay in which a proton is transformed into a neutron or vice versa, inside the nucleus of atom, process allowing the atom to move closer to the optimal ratio of protons and neutrons. As a result of this transformation, the nucleus emits a detectable beta particle, which is an electron or a positron.

There are two types of beta decay:

* *Beta plus Decay β⁺ or Positron Decay* - A proton is transformed into a neutron by losing a positively charged positron (beta plus particle) and a neutrino: $p \Rightarrow n + e^+ + v$

Positron decay or beta plus decay initiates the Sun's nuclear fusion, in which the proton in nucleus of element Hydrogen $^1H_1$ atom is transformed into a neutron, which then fuses with another Hydrogen atom under the Sun's conditions of extreme temperature and pressure to form Deuterium $^2H_1$ an isotope of element Hydrogen; the Deuterium nuclei undergo further fusion reactions to produce the element Helium $^4He_2$ with the release of huge amounts of energy.

* *Beta minus Decay β⁻ or Electron Decay* - A neutron is transformed into a proton by losing a negatively charged electron (beta minus particle) and an antineutrino: $n \Rightarrow p + e^- + \underline{v}$

Electron decay or beta minus decay is the basis of Radiocarbon dating.

In 1947 year, the American physical chemist **Willard Frank Libby** (1908-1980) published a paper outlining the principles of Radiocarbon dating, which revolutionized the archaeology. Willard Frank Libby received the **1960 Chemistry Nobel Prize** for his method to use carbon-14 for age determination in archaeology, geology, geophysics and other branches of science.

The element Carbon occurs naturally on Earth mainly in 3 isotopes:

two stable isotopes, carbon-12 ($^{12}C_6$) occurrence 99% and carbon-13 ($^{13}C_6$) occurrence 1%, and the radioactive isotope, carbon-14 ($^{14}C_6$) or radiocarbon in trace amounts.

Though it decays with a half life of about 5,730 years, the radioactive isotope of Carbon ($^{14}C_6$) is constantly being replenished by the activity of cosmic rays which react with the stable element Nitrogen-14 ($^{14}N_7$) atoms in the stratosphere and troposphere, transforming one of the Nitrogen's protons into a neutron, so creating the unstable radioactive isotope of Carbon ($^{14}C_6$), a cosmogenic nuclide.

This radiocarbon isotope quickly combines with the Oxygen in the atmosphere and forms Carbon dioxide CO_2, which is captured via photosynthesis by plants eaten by animals and so the radiocarbon is spread throughout the biosphere. During their lifetimes, plants and animals are constantly exchanging Carbon with their surroundings, so that the Carbon they contain will have the same proportion of radiocarbon, the carbon-14 ($^{14}C_6$) as the atmosphere.

Once the organism dies however, it ceases to acquire radiocarbon from the biosphere and the radiocarbon, carbon-14 ($^{14}C_6$) existing within its biological material at that time will decay with a half life of 5,730 years into Nitrogen-14 ($^{14}N_7$) transforming one of the Carbon's neutrons into a proton by *beta minus* or *electron decay*. The proportion of radiocarbon in the dead organism in comparison with its proportion in atmosphere can be used to determine how long it has been

since the organism died and stopped absorbing or ingesting radiocarbon, the carbon-14 ($^{14}C_6$).
The older the sample, the lower the percentage of carbon-14 ($^{14}C_6$) left in the dead organism.
The weak interaction is mediated by elementary/fundamental force carrier particles, which had
been found in experiments by the year 1983, the *gauge bosons W^+, W^-, Z.*

There are two types of weak interaction called "vertices":

- "charged-current interaction" mediated by particles that carry an electric charge (W^+ has +1e
electric charge and W^- has -1e electric charge) responsible for the beta decay phenomenon.
W bosons are named after the Weak nuclear force.

- "neutral-current interaction" mediated by a neutral particle, the Z boson. Z boson maybe is
so-named because it has Zero electric charge.

The weak interaction is the interaction responsible for all processes in which the particles' flavour
changes, hence for the instability of heavy quarks and leptons and particles that contain them.
Weak interactions that do not change particles' flavour have also been observed.

The decay of hadrons, like protons and neutrons, by the weak interaction can be viewed as
a process of decay of their constituent quarks. There is a pattern of these quark decays:
a quark of electric charge +2/3e (u,c,t) is always transformed to a quark of electric charge -1/3e
(d,s,b) and vice versa.

The decay of the up quark u in the down quark d, is important in the proton -> neutron
transformation. The decay of the down quark d in the up quark u, is important in the
neutron -> proton transformation.

The masses of W^+, W^- and Z bosons of 80-90 GeV/c^2 are far greater than that of protons or
neutrons. W^+, W^- and Z bosons have short lifetime under 10^{-24}s. The "theory of the weak
interaction" is sometimes called *quantum flavour-dynamics QFD* in analogy with the terms
quantum chromo-dynamics QCD and *quantum electrodynamics* QED,
but the term is rarely used because the weak force is best understood in terms of
"electro-weak theory" EWT.

2. Electromagnetism, type of physical field interaction occurring between electrically charged
particles, is the mechanism responsible for the *electromagnetic force*, the manifestation of
electromagnetic fields such as electric fields, magnetic fields and light.

The electromagnetic interaction is mediated by elementary/fundamental force carrier particles
photons, which are bosons. The photon has the symbol "γ".

The description of electromagnetic forces in terms of photons is a consequence of quantum theory
first defined by Paul Dirac.

In the paper "The quantum theory of the emission and absorption of radiation " appeared in

1927 year the scientist Paul Dirac does the first formulation of a quantum theory describing radiation and matter interaction.

Dirac described the quantization of the electromagnetic field as an ensemble of harmonic oscillators.

The quantum counterpart of classical electromagnetism, the relativistic quantum field theory of electrodynamics, giving a complete account of matter and light interaction is called *quantum electrodynamics QED*. The development of the QED theory was the basis of the **1965 Nobel Prize Physics**, awarded to physicists Americans **Richard Feynman** (1918-1988), **Julian Schwinger** (1918-1994) and Japanese **Shin-Ichiro Tomonaga** (1906-1979).

So precise and powerful is QED that it became the template for subsequent theories of other fundamental forces, such as weak and strong forces. Following the spectacular success of the *quantum electrodynamics* in the 1950s, attempts were undertaken to unify the electromagnetic force and the weak nuclear force by showing them to be two aspects of a single force, now termed the "electro-weak force". This culminated around 1968 year in the "electroweak theory" EWT, unified theory of electromagnetism and weak interactions by theoretical physicists Americans **Sheldon Glashow, Steven Weinberg** and Pakistani **Abdus Salam** (1926-1996) rewarded by the **1979 Nobel Prize in Physics**. The electroweak theory postulated W & Z bosons.

3. Strong interaction is the mechanism responsible for the *strong force* also called *strong nuclear force* or *colour force*.

The strong nuclear force is one of the four fundamental forces in nature, beside weak force, electromagnetic force and gravitation force and, as its name implies, is the *strongest* of the four. The strong force was first proposed to explain why atomic nuclei do not fly apart. It seemed that they would do so because of the repulsive electromagnetic force between the positively charged protons located in the nucleus. It was later found that the strong force main function is actually to bind together the quarks that make up the protons and neutrons. The force that holds nuclei together, was renamed the *residual strong force* or *residual colour force*.

The *strong nuclear force* is responsible for binding together the fundamental or elementary particles of matter to form larger particles, as it confines the quark elementary particles into hadron particles such as the proton and the neutron. It ensures stability for ordinary matter. The strong force holds the constituent quarks of a hadron together and is observable at less than 0.8 fm, which is the radius of a nucleon (proton or neutron).

The strong interaction is mediated by elementary/fundamental force carrier particles *gluons*, which are massless bosons. They are the "glue" that holds particles together.

The *residual strong force* holds hadrons, like protons and neutrons, together in nucleus and is observable at about 1-3 fm. The force carriers mediating the residual strong interaction are the particles called *mesons*.

The residual strong force drops off quickly at short distances and is only significant between adjacent particles within the nucleus.

The repulsive electromagnetic force, however, drops off more slowly, so it acts across the entire nucleus. Therefore, in heavy nuclei, particularly those with atomic numbers Z greater than 82 (lead), while the residual strong force on a particle remains nearly constant, the total electromagnetic force on that particle increases with atomic number Z (number of protons) to the point that eventually it can split the nucleus of atom and produce a large amount of energy or cause a large explosion, which is nuclear fission caused by the excess of the repulsive electrostatic force in comparison with the attractive residual strong force.

In nuclear fission reactions, electrostatic repulsion wins. The energy that is released by breaking the residual strong force bond takes the form of new high-speed particles and gamma rays, producing what is called radioactivity. Collisions with particles from the decay of nearby nuclei can precipitate this process causing a "nuclear chain reaction." Energy from the fission of heavy nuclei as those of uranium-235 and plutonium-239 powers nuclear reactors and atomic bombs.

If hadrons are struck by high-energy particles, they give rise to jets of massive particles, new hadrons, instead emitting free gluons because of the *colour confinement* that prevents the free "emission" of strong force. Particles affected by colour charge, like gluons and quarks, are subject to colour confinement and cannot exist free.

The "theory of strong interaction" is sometimes called *quantum chromo-dynamics QCD* and acquired its modern form around 1973–1974 years, when experiments confirmed that the *hadrons* were composed of fractionally electric charged *quarks*. The physicists **Sheldon Glashow** and **Howard Mason Georgi III** even proposed to unite electroweak and strong interaction fields in a "grand unified theory" GUT.

The particles that interact through the strongest force organize themselves into bound states on the smallest scales. Quarks and gluons, which couple to the strong force, typically bind into nuclear matter that can take myriad of forms like nuclei, protons, neutrons or quark-gluon plasma, depending on the pressures and temperatures involved. Larger systems, such as atoms and molecules built from these, are usually bound through the next-strongest interaction, electromagnetism.

4. Gravitation, type of physical field interaction, is the mechanism responsible for the *force of gravitation, gravitational force* or *gravity*, a natural phenomenon by which all physical bodies attract each other, an attractive force existing between any two objects that have mass.

The force of gravitation pulls objects together.

Since gravitational force is happening to all objects in the universe, from the largest galaxies down to the smallest atoms, the gravitation is often called universal gravitation.

The English physicist, mathematician, natural philosopher Sir Isaac Newton was the first to recognize that the force holding any object to the earth is the same as the force holding the moon, the planets and other heavenly bodies in their orbits.

According to Newton's law of universal gravitation, any two masses m and M with distance R between the centre of their masses, attract each other in universe with a gravitational force F given by the relation: $F = GMm/R^2$ G is the universal constant of gravitation

$$G = 6.7 \times 10^{-11} \, Nm^2/kg^2$$

The force unit was named *newton* (N) after Sir Isaac Newton.

The gravitational force is actually a very weak force. The attraction is too weak to be felt between two people or two objects placed right next to each other. It is only when one of the masses is the size of a planet that it is felt. The measure of gravitational force of an object is the weight of that object, which depends on the location of object in universe.

Approximately 10^{-38} times the strength of the strong force (38 orders of magnitude weaker), 10^{-36} times the strength of the electromagnetic force and 10^{-29} times the strength of the weak force, the gravitational force is the weakest of the four fundamental forces of nature and so it has a negligible influence on the behaviour of sub-atomic particles.

But the gravity is the dominant force at macroscopic scale, causing formation/evolution of astronomical bodies, galaxies, solar system, Earth and Moon & other phenomena observed on Earth and throughout the universe. That is because the gravitational force is the only force acting on all particles, it has an infinite range, it is always attractive and never repulsive and it cannot be absorbed, transformed or shielded against.

Electromagnetism is far stronger than gravitation, but the electromagnetism is not relevant to astronomical objects since such bodies have an equal number of protons and electrons resulting in zero electric charge.

The gravitational force is modelled as a continuous classical field while each of the other three fundamental forces is modelled as a discrete quantum field exhibiting elementary particles. It is possible to describe the gravity in the framework of quantum field theory like the other fundamental forces, like the attractive gravitational force arises due to exchange of "gravitons", in the same way as the electromagnetic force arises from exchange of *photons*, yielding acceptance of "quantum gravity theory" QGT.

On 14 September 2015 LIGO and Virgo Collaboration detected the first gravitational waves from a pair of merging black holes. The **2017 Physics Nobel Prize** was awarded to the physicists, German **Rainer Weiss** and Americans **Barry Clark Barish** and **Kip Stephen Thorne** for their contribution to the LIGO detector and observation of gravitational waves.

Remain widely disputed also phenomena suitable to model as a fifth force - perhaps an added gravitational effect.

All four fundamental interactions widely appear to align at an extremely minuscule scale, although the particle accelerators cannot produce the massive energy levels required to experimentally confirm that (Planck scale 1.22×10^{19} GeV).

The *string theory* ST, formulated in the 1970s years, was the first to describe strings of energy binding quark and antiquark to form a meson and according to it all particles in "zoo" emerge as vibrations of these strings. ST and other theories put both QGT and GUT within one framework, unifying all four fundamental interactions, within the "theory of everything" ToE. In pursuit of ToE, QGT has become an area of active research: it is hypothesized that the gravitational force is mediated by a massless boson the "graviton" with spin quantum number 2.

THE FOUR FUNDAMENTAL FORCES OF NATURE

FORCE		ACTS ON	AFFECTS PARTICLE	MEDIATOR
Gravitation		Mass-Energy	All particles	Graviton
Weak	Electroweak	Flavour	Quarks Leptons	$W^+W^-Z^0$
Electromagnetic		Electric charge	Electrically charged	Photon
Strong	Fundamental	Colour charge	Quarks Gluons	Gluon
	Residual	Atomic nuclei	Hadrons	Meson

The galaxies have a greater gravitational effect than the visible matter can explain, the "missing mass" problem. The physicists hypothesized the existence of "dark matter" to explain that. Dark matter has vastly more mass than the "visible" part of universe. Only 4% of the universe can be seen directly. 96% of universe is not visible, composed of 22% dark matter and 74% dark energy.

The current theory holds that the dark matter is composed of "weakly interacting massive particles" WIMPs that interact with ordinary matter only through gravity and weak interaction.

Invisible WIMPs with thousands of times the mass of a proton, have been hypothesized as being the substance of dark matter.

As the Sun, with its solar system, orbits around the centre of Milky Way galaxy, they pass through a halo of dark matter thought to be present in galaxy. This movement should generate a constant dark-matter "wind" on our planet, a stream of particles, which scientists are hoping to capture.

To directly detect strange stuff in the universe such as dark matter, there are dark-matter-detection projects operating on the same principle: Xenon Project in Italy, Large Underground Xenon Project (LUX) in South Dakota USA, Particle and Astrophysical Xenon Project (PandaX) in Sichuan China.

They start collect data in hopes to find evidence of the elusive particles, thought to constitute more than 80 precent of the matter in universe.

FERMIONS AND BOSONS

Universe is defined as the aggregate of all matter, energy, space and all its contents and time.

The Universe includes all, from smallest subatomic particles to biggest astronomic bodies.

The size of the whole Universe is not known and may be infinite.

The observable Universe is about 28.5 billion parsecs (93 billion light-years) in diameter at the present time. Observations and the development of physical theories have led to conclusions about the composition and evolution of the Universe.

The Universe is believed to consist of *Matter* and *Energy*.

Matter is anything that has mass and occupies space: solids, liquids, gases.

Energy is a property of objects, capacity of matter and radiation to perform work, it can be transferred or converted but cannot be created or destroyed.

Energy E is regarded to be inter-convertible with matter/mass m by the famous equation: $E=mc^2$

In Particle Physics, *elementary particle* or *fundamental particle* is a particle whose substructure is unknown, thus it is unknown whether it is composed of other particles.

A particle containing two or more elementary particles is a *composite particle*.

Known elementary particles include the fundamental "fermions" which generally are "matter particles" and "antimatter particles", as well as the fundamental "bosons" which generally are "force particles" that mediate interactions among fermions.

Properties of elementary particles include mass, spin, electric charge and colour charge and the different varieties of an elementary/fundamental particle are its flavours.

In Universe, the elementary/fundamental particles that make up, build the Matter are the *leptons* and the *quarks*. Each smaller building bloc of matter that was discovered was found by experiments that probed deeply into matter, but recent experiments have found no deeper structure than leptons and quarks.

Lepton is an elementary/fundamental particle in the hierarchy of Universe and one fundamental constituent of Matter. Leptons have various intrinsic properties, including mass, spin quantum number and electric charge.

Function of their electric charge, -1e or 0 e, there are two main types of leptons:

- *electric charged leptons* or *electron-like leptons* with three generations: electron, muon, tau

- *electric neutral leptons* or *neutrinos* with three generations: electron neutrino, muon neutrino, tau neutrino

So there are six flavours of leptons and for each of these leptons there is an *antilepton*. Between generations, the particles differ by their flavour quantum number and mass but their interactions are identical. Charged leptons can combine with other particles to form various composite particles such as *atoms* and *positronium*, while neutral leptons, neutrinos, rarely interact with anything and are consequently rarely observed. Leptons have half-integer spin quantum number s 1/2 and are subject to exclusion principle. Leptons are subject to weak interaction, gravitation and electromagnetism, the latest by excluding neutrinos, electrically neutral.

The best known of all leptons is the *electron,* which governs nearly all chemistry as it is found in atoms and is directly tied to all chemical properties. The *muon,* 207 times more massive than the electron with much greater mass of 105.7 MeV/c^2, is an unstable lepton with a mean lifetime of 2.2 μs, with negative electric charge of -1e and spin quantum number of 1/2 as the electron. The *tau* lepton has a lifetime of 2.9×10^{-13} s and a mass of 1776.82 MeV/c^2 (heaviest if compared to 105.7 MeV/c^2 for muons and 0.511 MeV/c^2 for electrons). Since its interactions are very similar to those of the electron, a tau can be thought of as a much heavier version of the electron. All three *neutrinos* have mass but that their masses are tiny, even by the standards of subatomic particles. Indications are that the summed masses of the three neutrinos must be less than 0.3 eV.

Quark is an elementary/fundamental particle in the hierarchy of Universe and the other fundamental constituent of Matter.

Function of their electric charge, there are two main types of quarks:

- *up-type quarks (electric charge +2/3)* with three generations: up, charm, top

- *down-type quarks (electric charge -1/3)* with three generations: down, strange, bottom

Quarks are the only known particles whose electric charges are not integer multiples of the elementary charge *e*. Quarks come in six flavours in terms of three pairs:

up/down, charm/strange and *top/bottom.*

The up and down quarks are stable and make up protons and neutrons. For example, the proton is composed of two *up quarks* and one *down quark* and is denoted as "uud".

The others, more massive flavours are only produced in high-energy interactions and have extremely short half-lives, typically observed in mesons.

Quarks have various intrinsic properties including mass, spin quantum number, electric charge and colour charge. Quarks have half-integer spin quantum number 1/2 and are subject to exclusion principle. Their property *colour charge* has three manifestations: *red, blue, green.*

Any quark with red, green or blue charge has a corresponding anti-quark whose colour charge must be the anti-colour of red, green or blue respectively *anti-red, anti-green* or *anti-blue,*

for the colour charge to be conserved in particle-antiparticle creation and annihilation.

All three colours mixed together, or any colour mixed with its anti-colour has zero colour charge, is "colourless" or "white".

The colour charged particles cannot exist independently because of their property called *colour confinement.* Follows that colour charged particles cannot be isolated singularly and therefore cannot be directly observed. By default, quarks clump together to form groups or hadrons, which are colourless. If a group is broken each fragment carries some of the colour charge and in order to obey the confinement, these fragments create other coloured particles around them so that the ensemble of these particles is a colourless jet. Jets are measured in particle detectors and studied in order to determine the properties of the original quarks.

Between quarks exist only two kinds of combinations: quark with antiquark and three quarks. Hundreds results are possible, all are particles that either have been discovered or are ready to be discovered. Quarks are the only fundamental/elementary particles to experience all four fundamental interactions.

Quarks combine to form composite particles called "hadrons".

Hadrons are defined simply as particles composed of quarks held together by the strong force; they are "quark-based" particles and have zero colour charge.

Hadrons are divided in *baryons* (composite of three quarks) and *mesons* (composite of one quark and one antiquark).

Baryon is a composite subatomic particle made up of three quarks bound together by the strong interaction. The word "baryon" comes from Greek word "βαρύς" (barys) meaning "heavy", because at the time of naming most known particles had lower masses than baryons.

```
                          Matter
                       /           \
                Hadrons              Leptons
              /         \
         Baryons          Mesons
        /    |    \       /     \
   quark quark quark   quark  antiquark
```

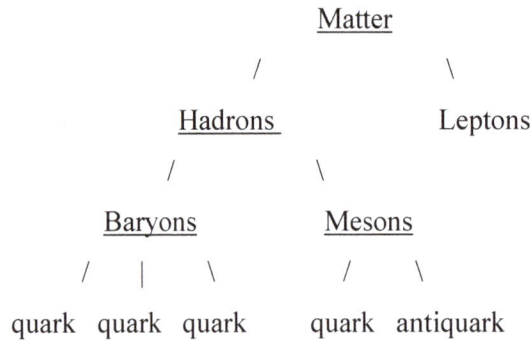

The most familiar baryons are the *protons* and *neutrons*. Protons and neutrons are key building blocks of matter and they make up most of the mass of the visible matter in the universe.

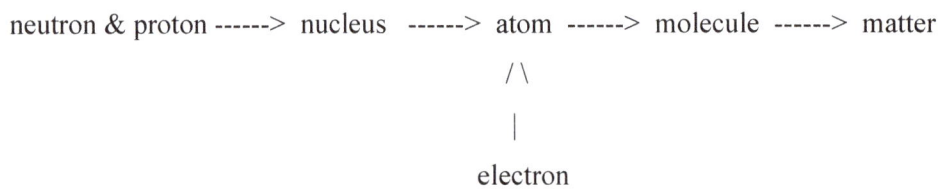

```
neutron & proton ------> nucleus ------> atom ------> molecule ------> matter
                                          / \
                                           |
                                        electron
```

Proton is a baryon, a composite subatomic particle made up of three quarks, two up quarks u and one down quark d, bound together by gluons. The colour assignment of individual quarks is arbitrary but all three colours must be present. The proton has zero total colour charge, is "colourless" or "white".

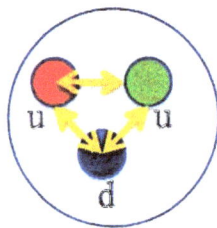

Proton, composed of two *up quarks u* and one *down quark d:* structure *uud.* Forces between quarks mediated by gluons

Proton has a physical size with a diameter roughly 1.6 fm, mass 938.2723 MeV/c^2, spin quantum number 1/2 and an electric charge of +1e. Considered for long time stable particles, recent development of grand unification models have suggested that a proton might decay with a lifetime of about 10^{32} years.

Neutron is a baryon, a composite subatomic particle made up of three quarks, one up quark u and two down quarks d, bound together by gluons. The colour assignment of individual quarks is arbitrary but all three colours must be present. The neutron has zero total colour charge, is "colourless" or "white".

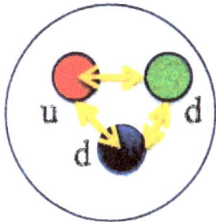

Neutron, composed of one *up quark* and two *down quarks*:
structure *udd.* The quarks bounded by *gluons*

Neutron has a physical size slightly bigger than proton, mass 939.5656 MeV/c^2 (0.2% more massive than proton) spin quntum number 1/2 and no electric charge. A free neutron will decay with a lifetime just under 15 minutes but it is stable if combined in nucleus.

Meson is a composite subatomic particle made up of one quark and one antiquark, bound together by the strong interaction carrier.

The quark can be any colour and the antiquark will have the negative (complement) of that colour. The meson has zero total colour charge, is "colourless" or "white".

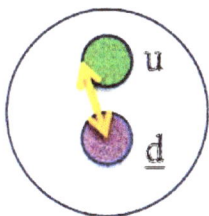

Meson, composed of one *quark* and one *antiquark* bounded by gluon

Mesons have a physical size with a diameter roughly 1fm, which is about 2/3 the size of a proton or neutron, mass between 139MeV/c^2 and 9.460 GeV/c^2, spin quntum number 0 or 1 and electric charge of -1e, 0e, +1e.

All meson types arc unstable, with the longest-lived lasting for only a few hundredths of a microsecond. Charged mesons decay to form electrons and neutrinos. Uncharged mesons may decay to photons.

Each type of meson has a corresponding antiparticle (anti-meson) in which quarks are replaced by their corresponding antiquarks and vice versa. For example, the positive pi-meson, *pion π+*, is made of one *up quark* and one *down antiquark*; and its corresponding antiparticle, the negative pi-meson, *pion π−*, is made of one *up antiquark* and one *down quark*.

The colour properties explain how the quarks are able to obey the *exclusion principle*, which states that no two identical objects can occupy the same place, meaning quarks making up the same hadron must have different colours.

In quantum chromodynamics, the modern theory of the nuclear force, most of the mass of protons, neutrons and mesons is explained by special relativity.

For example, the mass of a proton is about 80–100 times greater than the sum of the masses of the quarks that make it up, while the gluons have zero mass. The extra energy of the quarks and gluons within the proton, as compared to the energy of the quarks alone, accounts for almost 99% of the mass. The mass of a proton is, thus, the mass of the system of moving quarks and gluons that make up the proton.

The leptons and quarks belong to the class of particles called *fermions*.

The fermions can be elementary particles as the electrons or composite particles as the protons; they are named *fermions* after the statistics in Quantum Mechanics that correctly describe their behaviour, the "Fermi-Dirac statistics".

Fermi-Dirac statistics (F-D statistics) is one of two possible ways in which a system of indistinguishable particles with negligible mutual interaction can be distributed among a set of energy states: *each of the available discrete energy states can be occupied by only one particle.* The theory of this statistical behaviour, was discovered in the 1926–1927 years by the physicists Enrico Fermi and Paul Dirac independently.

The Fermi-Dirac statistics apply only to those types of particles that have half-integer values of spin quantum number in a system in thermodynamic equilibrium, obeying the restriction, known as the *Pauli principle* or *exclusion principle*, that no two identical particles can occupy the same energy state. This allows the many-particle system to be described in terms of single-particle energy states. It is most commonly applied to electrons, which are fermions with spin quantum number 1/2. The dynamics of objects with spin quantum number 1/2 cannot be accurately described using classical physics; they are among the simplest systems, which require Quantum Mechanics to describe them.

The fermions include the elementary/fundamental particles *quarks* and *leptons*, as well as any composite particle made of an odd number of those, as all *baryons* and many *atoms* and *nuclei*.

The particles that do not obey the *Pauli principle* or *exclusion principle* restrictions belong to the class of particles named *bosons*, after the statistics that correctly describe their behaviour, the "Bose–Einstein statistics".

Bose–Einstein statistics (B–E statistics) is one of two possible ways in which a collection of non-interacting indistinguishable particles may occupy a set of available discrete energy states at thermodynamic equilibrium: *those particles are not limited to single occupancy of the same energy state.*

The theory of this statistical behaviour was developed in the 1924–1925 years by the Indian mathematical physicist **Satyendra Nath Bose** (1894-1974), who recognized that a collection of identical and indistinguishable particles can be distributed in this way, idea later adopted and extended by Albert Einstein in collaboration with Satyendra Nath Bose.

The Bose–Einstein statistics apply only to those types of particles that have integer values of the spin quantum number in a system in thermodynamic equilibrium, not obeying the restriction known as the *Pauli principle* or *exclusion principle*.

Two or more identical boson particles can occupy the same energy state, can exist in the same place at the same time.

The bosons may be either elementary/fundamental particles or composite particles:

       * elementary/fundamental particles

gauge bosons:

- *photon* γ force carrier of the electromagnetic force

- *gluon* g force carrier of the strong force

- *W bosons* W^+ W^- and *Z boson* Z^0 force carriers of the weak force

- *graviton* G which is still-theoretical force carrier for gravity

scalar bosons:

- *Higgs boson* H^0 which gives other particles their mass via the Higgs mechanism

       * composite particles

- *mesons*

- *stable nuclei of even mass number* such as *lead-208, helium-4, deuterium* (A= 2)

       * quasiparticles

- *Cooper pairs*

- *plasmons*

- *phonons*

Forces are carried by bosons with non-zero spin, the "gauge bosons".

Gauge boson is a force carrier, one that mediates any of the fundamental interactions of nature.

Spin-statistics theorem of Quantum Mechanics, relates the spin of particle to its statistics:

- Particles of *half-integer* spin quantum number exhibit Fermi–Dirac statistics and are fermions

- Particles of *integer* spin quantum number exhibit Bose–Einstein statistics and are bosons

The Spin-statistic theorem underlines the important characteristic that bosons statistics does not restrict the number of particles that occupy the same quantum state of energy, whereas fermions statistics restricts the number of particles that occupy the same quantum state of energy at only one particle.

The two classes of particles known as *Fermions* and *Bosons* are the two kinds of things possible in universe and the dialectic between them describes all physical form. Fermions are particles usually associated with **Matter** whereas bosons are generally **Force** carrier particles.

ELEMENTARY PARTICLES

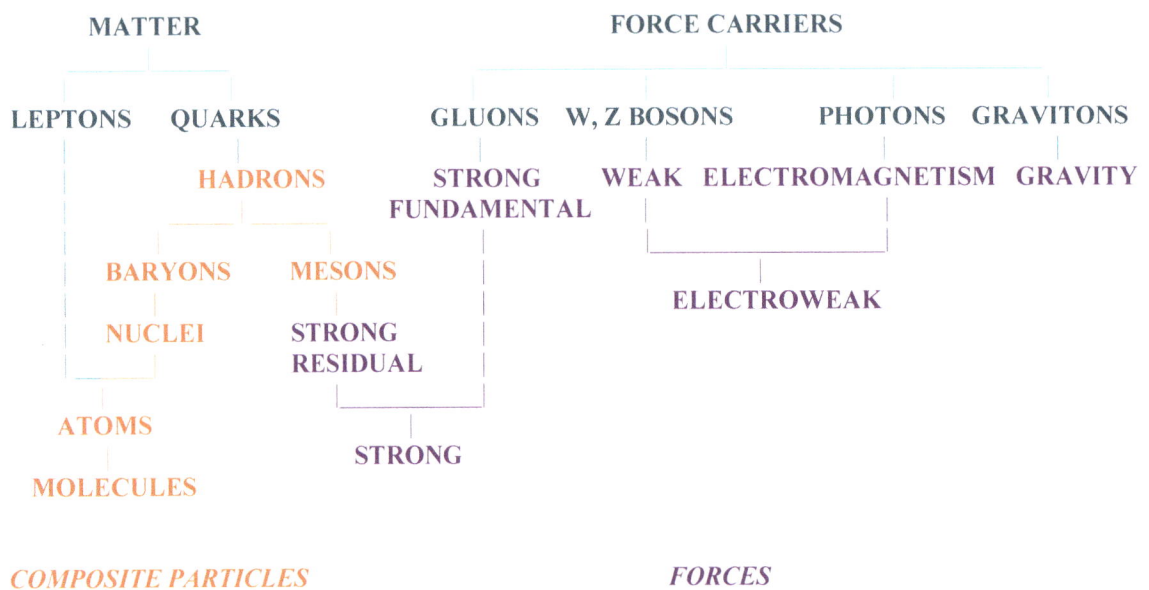

MATTER **FORCE CARRIERS**

LEPTONS QUARKS   **GLUONS W, Z BOSONS   PHOTONS GRAVITONS**

HADRONS   **STRONG   WEAK ELECTROMAGNETISM GRAVITY**
FUNDAMENTAL

BARYONS   MESONS

**NUCLEI   STRONG
RESIDUAL**   **ELECTROWEAK**

ATOMS

MOLECULES   **STRONG**

COMPOSITE PARTICLES   *FORCES*

STANDARD MODEL

In Particle Physics, the reigning theory is the **Standard Model**, which describes the basic building blocks of matter, the smallest particles of matter and how they interact.

The Standard Model is a quantum field theory, based on two quantum field theories:

- *Quantum electrodynamics* QED, based on electromagnetic force quanta, which explains how electrons, positrons and photons interact.

- *Quantum chromodynamics* QCD, based on strong force quanta, which explains how quarks and gluons interact.

The Standard Model describes the Universe in terms of **Matter** (fermions) and **Force** (bosons). In addition to all the known and predicted subatomic particles, the Standard Model includes the strong force, weak force and electromagnetism and explains how these forces act on the particles of matter. It does not include the gravitational force. According to CERN, at the scale of subatomic particles, the effect of gravitation is so minuscule that the model works well despite its exclusion.

According to the Standard Model all mass consists of fermions. How the fermions combine to form objects or do not combine at all depend on the fundamental forces.

According to the Standard Model all forces are mediated by *gauge* bosons:

W+, W–, Z bosons for weak force, *photons* for electromagnetic force, *gluons* for strong force.

Under the Standard Model, one of the smallest elementary/fundamental particles that cannot be split up into smaller parts is the *quark*. The quarks are the building blocks of the family of massive particles known as *hadrons*, which includes *protons* and *neutrons*. Most of the Matter we see around us is made from protons and neutrons. Scientists haven't seen any indication that there is anything smaller than a quark, but they still search.

Upon experimental confirmation of the existence of *quarks* in 1968 year (theorized in 1964), the current formulation of the Standard Model was finalized in the mid-1970s.

The discoveries of *W & Z bosons* in 1981 year, *top quark* in 1995 year, *tau neutrino* in 2000 year and more recently *Higgs boson* in 2013 year, after they were predicted, have made the Standard Model more convincing.

Mathematically, the Standard Model is a quantized *Yang–Mills theory*. Yang–Mills theory seeks to describe the behaviour of elementary particles using non-commutative Lie-groups, which represent the best developed theory of continuous symmetry of mathematical objects and structures, which makes them indispensable tools for many parts of contemporary mathematics, as well as for modern theoretical physics.

The Standard Model is a theory concerning the fundamental interactions, which mediate the dynamics of the known subatomic particles; also it is a paradigm of a quantum field theory, which exhibits a wide range of physics including spontaneous symmetry breaking, anomalies, non-perturbative behaviour, etc.

Result of an inspired theoretical effort, the Standard Model of Particle Physics was developed further as a collaborative effort of both theoretical and experimental particle physicists.

The Standard Model is used as a basis for building more *exotic models* that incorporate hypothetical particles, extra dimensions and elaborate symmetries such as super-symmetry, in an attempt to explain experimental results at variance with it, such as the existence of dark matter and neutrino oscillations.

All particles and their interactions, observed to date, can be described almost entirely by the quantum field theory of the Standard Model.

Antimatter (fermions) is being regarded as a kind of *Matter* (fermions).

Because of its success in explaining a wide variety of experimental results, the Standard Model is sometimes regarded as a "theory of almost everything". With the construction of the Large Hadron Collider LHC at CERN, the Standard Model will continue to be a vital and active subject.

The Standard Model describes approximately 200 particles and their interactions using 61 elementary/fundamental particles, all of which are fermions or bosons:

18 quarks + 18 anti-quarks, 6 leptons + 6 anti-leptons, 8 gluons, 1 W+ boson, 1 W– boson, 1 Z boson, 1 photon, 1 Higgs boson

Those elementary particles can combine to form composite particles, accounting for the hundreds species of particles that have been discovered since 1960s.

The Standard Model has demonstrated huge and continued successes in providing predictions of experiments and found to agree with almost all experimental tests conducted to date.

However about 85% of the matter in Universe is yet unaccounted for by any of the particles in the Standard Model. It is the missing "dark matter".

And the Standard Model cannot explain some phenomena, as in recent years measurements of neutrino mass have provided the first experimental deviations from the Standard Model.

At present, the most particle physicists believe that the Standard Model is an incomplete description of Nature and that a more fundamental theory awaits to be discovered, the "theory of everything" ToE, involving the phenomenon of super-symmetry SUSY, which can be described in mathematical equations of an uncommon beauty.

Elementary Particles	Quark	Lepton	Gluon	W Boson	Z Boson	Photon	Higgs Boson
Types	2	2	1	1	1	1	1
Generations	3	3	1	1	1	1	1
Antiparticles	Yes	Yes	Own	Yes	Own	Own	Own
Colours	3	-	8	-	-	-	-
Total	36	12	8	2	1	1	1

NEW CHEMICAL ELEMENTS

The periodic table of chemical elements is considered by many the most comprehensive classification system in the world because it contains all the existing building blocks that compose the universe.

The history of the periodic table is also the history of the discovery of the chemical elements. After Marie Sklodowska Curie's discoveries in radioactivity, Mendeleev's periodic table of elements has been expanded and improved with the discovery or synthesis of new chemical elements and the development of new theoretical models to explain chemical behaviour.

At present, the periodic table of chemical elements has 118 elements from hydrogen with atomic number $Z=1$ to ununoctium with atomic number $Z=118$ and it is considering the synthesis of elements with higher atomic numbers.

references

MARIE SKLODOWSKA CURIE / HER CONTRIBUTION TO SCIENCE – IRINA RODICA RABEJA 2016

ANNEX

File:Wireless-icon.png

From Wikipedia, the free encyclopedia

Size of this preview: 750 × 600 pixels.
Original file (1,280 × 1,024 pixels, file size: 5.01 MB, MIME type: image/png)

pixabay

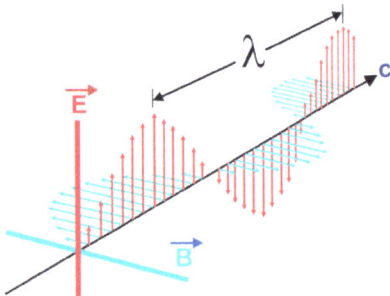

Creative Commons CC0 CC0 1.0 Universal (CC0 1.0)
Public Domain Dedication   No Copyright

File:Atmospheric electromagnetic opacity.svg

From Wikimedia Commons, the free media repository

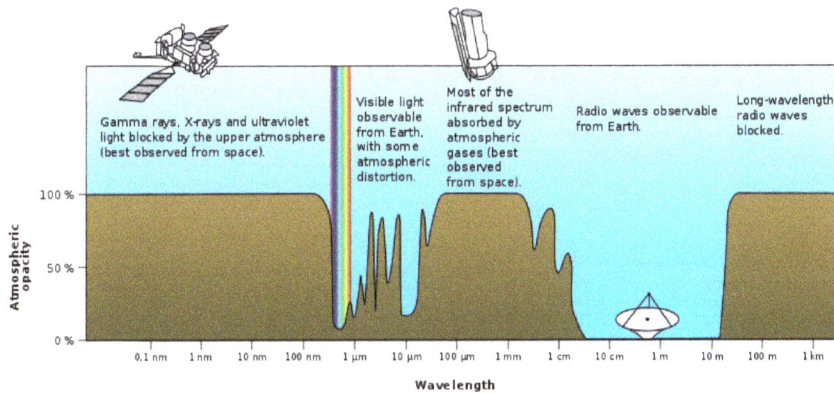

Size of this PNG preview of this SVG file: 800 × 378 pixels.
Original file (SVG file, nominally 1,650 × 780 pixels, file size: 59 KB)

This image rendered as PNG in other widths: 200px, 500px, 1000px, 2000px.

Open in Media ViewerConfiguration

Description	**English:** Electromagnetic transmittance, or opacity, of the Earth's atmosphere. **Magyar:** A Föld atmoszférájának elektromágneses transzmittanciája (más néven opacitása, azaz százalékos áteresztőképessége) a hullámhossz függvényében.
Date	27 December 2008 (SVG)
Source	Vectorized by User:Mysid in Inkscape, original NASA image from File:Atmospheric electromagnetic transmittance or opacity.jpg.
Author	NASA (original); SVG by Mysid.

Licensing

This file is in the *public domain* in the United States because it was solely created by *NASA*. NASA copyright policy states that "NASA material is not protected by copyright **unless noted**". (See *Template:PD-USGov, NASA copyright policy page* or *JPL Image Use Policy*.)

File:EM Spectrum Properties es.svg

From Wikimedia Commons, the free media repository

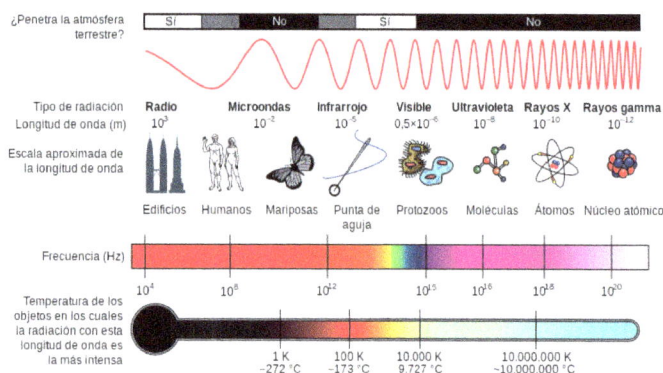

Size of this PNG preview of this SVG file: 700 × 400 pixels.
Original file (SVG file, nominally 700 × 400 pixels, file size: 185 KB)

Description	**Deutsch:** Ein Diagramm des elektromagnetischen Spektrums, mit Angabe der Benennung, Wellenlänge (mit Beispielen), Frequenz, Temperatur der entsprechenden Schwarzkörperstrahlung. In Anlehnung an eine Vorlage der NASA **English:** A diagram of the EM spectrum, showing the type, wavelength(with examples), frequency, the black body emission temperature. Temporary file for gauging response to an improved version of this file. Adapted from EM_Spectrum3-new.jpg, which is a NASA image. **Español:** Diagrama del espectro electromagnético, mostrando el tipo, longitud de onda (con ejemplos), frecuencia y la temperatura de emision de cuerpo negro. Imagen adaptada de esta otra de la NASA. **Français :** Diagramme montrant le spectre éléctromagnétique avec le type, la longueur d'onde (avec des exemples de tailles), la fréquence, et la température d'emission du corps noir. Image adaptée d'un document de la NASA.
Date	17 February 2008
Source	Translation from English version
Author	Crates. Original version in English by Inductiveload
Permission (Reusing this file)	I, the copyright holder of this work, release this work into the **public domain**. This applies worldwide. In some countries this may not be legally possible; if so: *I grant anyone the right to use this work **for any purpose**, without any conditions, unless such conditions are required by law.*

File:Faraday Cochran Pickersgill.jpg

From Wikimedia Commons, the free media repository

Size of this preview: 480 × 600 pixels.
Original file (1,155 × 1,443 pixels, file size: 486 KB, MIME type: image/jpeg)

Summary

Description	**English:** Michael Faraday, (22 September 1791 – 25 August 1867) in his late thirties
Date	circa 1826
Source	Dibner Library, http://ihm.nlm.nih.gov/images/B07069
Author	Painted by H.W. Pickersgill (1782-1875), Engraved by John Cochran (1821-1865)
Permission (Reusing this file)	Artists and sitter died over 140 years ago

Licensing

This is a faithful photographic reproduction of a two-dimensional, public domain work of art. The work of art itself is in the public domain for the following reason:

This work is in the public domain in its country of origin and other countries and areas where the copyright term is the author's **life plus 100 years or less**.

_____

⚠ You must also include a United States public domain tag to indicate why this work is in the public domain in the United States.
This file has been identified as being free of known restrictions under copyright law, including all related and neighboring rights.

139

File:James Clerk Maxwell.png

From Wikimedia Commons, the free media repository

James_Clerk_Maxwell.png (331 × 398 pixels, file size: 101 KB, MIME type: image/png)

Description	Engraving of James Clerk Maxwell by G. J. Stodart from a photograph by Fergus of Greenock
Date	Unknown date
Source	Frontpiece in James Maxwell, *The Scientific Papers of James Clerk Maxwell*. Ed: W. D. Niven. New York: Dover, 1890.
Author	**George J. Stodart**
Permission (Reusing this file)	As a work from sometime before 1890, in the public domain.

File:Heinrich Hertz.jpg

From Wikimedia Commons, the free media repository

Size of this preview: 463 × 599 pixels.
Original file (742 × 960 pixels, file size: 236 KB, MIME type: image/jpeg)

Description	**Deutsch:** Porträt Heinrich Hertz, Fotografie von Robert Krewaldt, Bonn Veröffentlich bzw. erstellt von "Washington, D.C.: Underwood & Underwood", nachbearbeitet / Lizenz: Public Domain
Date	1915; image first published 1894 at the latest
Source	http://memory.loc.gov
Author	Robert Krewaldt

File:Lindsay James Bowman.jpg

From Wikimedia Commons, the free media repository

[Lindsay_James_Bowman.jpg](#) (214 × 319 pixels, file size: 18 KB, MIME type: image/jpeg)

File:Abraham Archibald Anderson - Thomas Alva Edison - Google Art Project.jpg

From Wikimedia Commons, the free media repository

Size of this preview: 728 × 600 pixels.
Original file (5,778 × 4,761 pixels, file size: 4.24 MB, MIME type: image/jpeg);   Summary

Artist	Abraham Archibald Anderson (1847 - 1940) Details of artist on Google Art Project
Title	***Thomas Alva Edison***
Object type	Painting
Date	1890
Medium	oil on canvas
Dimensions	Height: 1,143 mm (45 in). Width: 1,387 mm (54.61 in).
Current location	National Portrait Gallery, Washington
Accession number	NPG.65.23
Notes	More info at museum site
Source/Photographer	3QEAJP0QlAZWiw at Google Cultural Institute maximum zoom level

Licensing

This is a faithful photographic reproduction of a two-dimensional, public domain work of art. The work of art itself is in the public domain for the following reason:

The author died in 1940, so this work is in the public domain in its country of origin and other countries and areas where the copyright term is the author's **life plus 75 years or less**.

This work is in the public domain in the United States because it was published (or registered with the U.S. Copyright Office) before January 1, 1923.

This file has been identified as being free of known restrictions under copyright law, including all related and neighboring rights.

The official position taken by the Wikimedia Foundation is that "*faithful reproductions of two-dimensional public domain works of art are public domain*".
This photographic reproduction is therefore also considered to be in the public domain in the United States. In other jurisdictions, re-use of this content may be restricted; see Reuse of PD-Art photographs **for details.**

File:N.Tesla.JPG

Size of this preview: 459 × 600 pixels.
Original file (2,563 × 3,348 pixels, file size: 1.4 MB, MIME type: image/jpeg);

Description	**English:** A photograph of Nikola Tesla (1856-1943) at age 40. **Polski:** Nikola Tesla (1856-1943), serbski inżynier i wynalazca, na zdjęciu w wieku lat 40 **Русский:** Фотография 40-летнего физика-изобретателя Никола Тесла (1856-1943). **Українська:** Фотографія 40-річного фізика-винахідника Нікола Тесла (1856-1943).
Date	circa 1896
Source	Downloaded from: https://historyrat.wordpress.com/2013/01/13/lighting-the-1893-worlds-fair-the-race-to-light-the-world/
Author	Unknown
Permission (Reusing this file)	This image is in the public domain due to its age; PD-OLD.

Licensing

File:TeslaWirelessPower1891.png   From Wikipedia, the free encyclopedia

TeslaWirelessPower1891.png (558 × 371 pixels, file size: 264 KB, MIME type: image/png)

This is a file from the Wikimedia Commons. Information from its description page there is shown below.
Commons is a freely licensed media file repository. You can help.

Description	**English:** Nikola Tesla demonstrating wireless transmission of power and high frequency energy at Columbia College, New York, in 1891. The two metal sheets were connected to his Tesla coil oscillator, which applied a high voltage oscillating at radio frequency. The electric field ionized the gas in the long partially-evacuated Geissler tubes he is holding (similar to modern neon lights), causing them to emit light without wires. **Interlingua:** Tesla demonstrante transmissiones sin filos durante su lection super alte frequentias e potential de 1891. Post ulterior recercas, Tesla presentava le fundamentos del radio in 1893. **Esperanto:** Nikola Tesla montras siajn eksperimentojn en Columbia College, Novjorko, en 1891. Post daŭraj esploroj li montris fundamentojn de radio en 1893.
Date	1891
Source	Nikola Tesla, "Experiments with alternate currents of very high frequency and their application to methods of artificial illumination" in *Electrical World* magazine, W. J. Johnson Co., New York, Vol. 18, No. 2, May 20, 1891, p. 19 Transferred from en.wikipedia to Commons.
Author	Nikola Tesla

Licensing

File:Tesla Broadcast Tower 1904.jpeg

From Wikipedia, the free encyclopedia

Tesla_Broadcast_Tower_1904.jpeg (472 × 512 pixels, file size: 61 KB, MIME type: image/jpeg)

This is a file from the Wikimedia Commons. Information from its description page there is shown below.
Commons is a freely licensed media file repository. You can help.

Summary

Description **English:** Nikola Tesla's Wardenclyffe wireless station, located in Shoreham, New York, seen in 1904. The 187 foot (57 m) transmitting tower appears to rise from the building but actually stands on the ground behind it. Built by Tesla from 1901 to 1904 with backing from Wall Street banker J. P. Morgan, the experimental facility was intended to be a transatlantic radiotelegraphy station and wireless power transmitter, but was never completed. The tower was torn down in 1916 but the lab building, designed by noted New York architect Stanford White remains.

Date 1904

Source
- Retrieved from http://www.sftesla.org/images/Tesla_Broadcast_Tower.JPG
- Previously published in Arthur B. Reeve, "Tesla and his Wireless Age" in *Popular Electricity* magazine, Popular Electricity Publishing Co., Chicago, Vol. 4, No. 2, June 1911, p. 97

Author Unknown (Life time: Unattributed)

Licensing

File:Guglielmo Marconi.jpg

From Wikimedia Commons, the free media repository

Size of this preview: 450 × 599 pixels.
Original file (987 × 1,314 pixels, file size: 196 KB, MIME type: image/jpeg)

Description	Guglielmo Marconi, portrait, head and shoulders, facing left.
Date	Copyright 1908
Source	This image is available from the United States Library of Congress's Prints and Photographs division under the digital ID cph.3a40043. This tag does not indicate the copyright status of the attached work. A normal copyright tag is still required. See Commons:Licensing for more information. العربية | čeština | Deutsch | English | español | فارسی | suomi | français | magyar | italiano | македонски | Nederlands | polski | português | русский | slovenčina | slovenščina | Türkçe | українська | 中文 | 中文（简体）| 中文（繁體）| +/−
Author	Pach Brothers
Permission (Reusing this file)	This work is in the **public domain** in the United States because it was published (or registered with the U.S. Copyright Office) before January 1, 1923. ⚠ Public domain works must be out of copyright in both the United States and in the source country of the work in order to be hosted on the Commons. If the work is not a U.S. work, the file **must** have an additional copyright tag indicating the copyright status in the source country.

File:NormanAbramson.jpg

From Wikimedia Commons, the free media repository

Size of this preview: 428 × 600 pixels.
Original file (731 × 1,024 pixels, file size: 423 KB, MIME type: image/jpeg)

Summary

Description	**English:** Picture of Norman Abramson, 2007.
Date	
Source	Norman Abramson
Author	Norman Abramson

Licensing

I, the copyright holder of this work, release this work into the public domain. This applies worldwide.
In some countries this may not be legally possible; if so:
*I grant anyone the right to use this work **for any purpose**, without any conditions, unless such conditions are required by law.*

File:Marie Curie 1903.jpg

From Wikimedia Commons, the free media repository

This file has been **extracted** from another file: Marie-noble-portrait-600.jpg.

Description	**English:** Marie Curie
Date	1903
Source	http://nobelprize.org/nobel_prizes/physics/laureates/1903/marie-curie-bio.html
Author	Nobel foundation

Permission
(Reusing this file)

File:Ernest Rutherford 1908.jpg

Original file (3,227 × 4,288 pixels, file size: 1.75 MB, MIME type: image/jpeg);

Summary

Description	Ernest Rutherford (1871-1937), nuclear physicist
Date	circa 1896 (1908-12-03 is unimpossible! this would meet the date of nobel prize)
Author	Bain News Service, publisher
Permission (Reusing this file)	*This work is from the George Grantham Bain collection at the Library of Congress. According to the library, there are no known copyright restrictions on the use of this work.*
Other versions	*This media file is in the public domain in the United States. This applies to U.S. works where the copyright has expired, often because its first publication occurred prior to January 1, 1923. See this page for further explanation.*

www.ingramcontent.com/pod-product-compliance
Lightning Source LLC
Chambersburg PA
CBHW041620220326
41597CB00035BA/6182